Modern Operations with Future Integrated Vehicle Management Systems 2021

Online
19 - 20 August 2021

ISBN: 978-1-7138-8786-7

Printed from e-media with permission by:

Curran Associates, Inc.
57 Morehouse Lane
Red Hook, NY 12571

CURRAN ASSOCIATES INC.
proceedings
.com

Some format issues inherent in the e-media version may also appear in this print version.

Vertical Flight Society
2700 Prosperity Ave, Suite 275
Fairfax, VA 22031
USA

Phone: (703) 684-6777
Fax: (703) 739-9279

www.vtol.org

Additional copies of this publication are available from:

Curran Associates, Inc.
57 Morehouse Lane
Red Hook, NY 12571 USA
Phone: 845-758-0400
Fax: 845-758-2633
Email: curran@proceedings.com
Web: www.proceedings.com

TABLE OF CONTENTS

Agenda

Vertical Flight Society Philadelphia Chapter

Vertical Flight Society Aberdeen Chapter

Modern Operations with Future Integrated Vehicle Management Systems
***VIRTUAL* Technical Meeting**

Thursday Aug. 19 – Friday Aug. 20, 2021

Thursday August 19, 2021
Virtual Presentations will take place in US Eastern Daylight Time – Time & Date

10:00 – 10:15 AM	**Welcome/Introductions**	
10:15 – 10:45 AM	Technical Paper Presentation *Piloted Evaluation of Adaptive Flight Controls Designed to Improve Performance, Safety, and Survivability for Advanced VTOL*	**Derek Bridges**, Piasecki Aircraft Corp.
10:45 – 11:15 AM	Technical Paper Presentation *Machine-learning Framework for Online Probabilistic Rotor Fault Detection, Identification and Quantification on a Multicopter using Boom Strain Signals*	**Airin Dutta**, RPI – Center for Mobility with Vertical Lift (MOVE)
11:15 – 11:45 AM	Invited Presentation *Vision of an ATM System for Future AAM*	**Nick Lappos**, Sr. Fellow, Sikorsky, LMCO & Chair, Nat'l. Academy of Eng., Aero. & Space Eng. Board
11:45 AM–12:15 PM	Keynote Presentation *Advanced VMS – A Perspective*	**Russell Enns and Fernando Dones**, Boeing Vertical Lift
12:15 – 12:20 PM	**Sponsor Presentation** ***Accelerating Avionics Design & Deployment*** ***Real-Time Innovations (RTI)***	**Ross Gilson**, Senior Field Applications Engineer, RTI
12:20 – 12:45 PM	**Break**	
12:45 – 1:15 PM	Technical Paper Presentation *Candidate VMS Architecture Approaches to Support Future AH-64 Apache Attack Helicopter Requirements*	**Gary Klein**, Boeing Vertical Lift
1:15 1:45 PM	**Technical Presentation** ***Domain-Specific Language Configuration for COTS Flight-Control Software Building Blocks***	**Ivan Makarov**, Catalyst Aerospace Technologies
1:45 – 2:15 PM	Technical Paper Presentation *Evolving System Integration Labs in the Age of Digital Transformation*	**Bill Eccles**, Bloomy Controls
2:15 – 2:30 PM	**Break**	
2:30 – 3:00 PM	Invited Presentation AIAA *Case Studies in Adaptive Flight Control*	**Dr. Anthony J. Calise**, Georgia Institute of Tech. (Ret.)
3:00 – 3:30 PM	Invited Presentation Near Earth Autonomy *Assured Autonomy: Why? What? How?*	**Sanjiv Singh**, Near Earth Autonomy
3:30 – 4:00 PM	Invited Presentation: Vertical Flight Society (VFS) *The Future of Vertical Flight*	**Mike Hirschberg**, Executive Director, VFS
4:00 – 4:15+ PM	**Day 1 Wrap-up**	

Agenda

Modern Operations with Future Integrated Vehicle Management Systems
VIRTUAL Technical Meeting
Thursday Aug. 19 – Friday Aug. 20, 2021

Vertical Flight Society Philadelphia Chapter
Vertical Flight Society Aberdeen Chapter

Friday August 20, 2021
Virtual Presentations will take place in US Eastern Daylight Time – Time & Date

10:00 – 10:15 AM	**Welcome/Introductions**	
10:15 AM – 12:00 PM	*Panel Discussion: Reimagining Vehicle Management Systems for UAS* **Matt Rhinehart, Flight Dynamics Branch Head, Naval Warfare Center Aircraft Division** **Ted Maciuoba, Deputy Director, Robotics Requirements Army Futures Command** **Dr. Bill Evans, Human Agent Teaming Essential Research Program DEVCOM-Army Research Lab** **Dr. Tom Berger, Flight Controls Lead, DEVCOM AvMC** **Mark Butkiewicz, V.P., SURVICE Engineering Co.**	**Moderated by Dr. Jean-Paul Reddinger**, DEVCOM Army Research Lab
12:00 PM – 12:15 PM	**Break**	
12:15 – 12:45 PM	Technical Presentation *Role of Physics-Based Models in Simulation for VMS*	**Michael Avera**, DEVCOM Army Research Lab
12:45 – 1:15 PM	Invited Technical Presentation *A GNC Architecture for Autonomous Transitioning Aerial Vehicles*	**Kristoff McIntosh and Prof. Sandipan Mishra** RPI – Center for Mobility with Vertical Lift (MOVE)
1:15 –1:45 PM	Invited Technical Presentation *Deterministic Reconfiguration of Flight Control Systems for Multirotor UAV Package Delivery*	**Anthony Gong,** DEVCOM AvMC
1:30 – 1:55 PM	Invited Technical Presentation *Coordinated Planning with Aerial and Ground Vehicles for Long-Term Autonomy*	**Prof. Pratap Tokekar,** University of Maryland
1:45 – 2:00 PM	**Meeting Wrap-up**	

General Chair: Mr. Dan Wells, Leonardo

Technical Co-Chairs:
Mr. Kevin W. Wise, Boeing Vertical Lift
Dr. Rajneesh Singh, DEVCOM Army Research Lab

Thanks to our sponsor Real-Time Innovations (RTI)

rti

Piloted Evaluation of Adaptive Flight Controls Designed to Improve Performance, Safety, and Survivability for Advanced VTOL Aircraft

Derek O. Bridges
Flight Controls Engineer
Piasecki Aircraft Corporation
Essington, PA, USA

Tom Berger
Flight Controls Group Lead
U.S. Army Combat Capabilities
Development Command
Aviation & Missile Center
Moffett Field, CA, USA

Joseph F. Horn
Professor
Department of Aerospace
Engineering
The Pennsylvania State University
University Park, PA, USA

D. Grey Hagwood, Jr.
Director of Advanced Concepts
Piasecki Aircraft Corporation
Essington, PA, USA

ABSTRACT

ADAPT™ is a flight control software package that improves the safety, survivability, performance, and affordability of advanced VTOL aircraft that include multiple types and instances of control effectors. ADAPT™ includes three different control algorithms: Fly-to-Optimal performs trim optimization based on selected objectives for increased performance; Vehicle Maneuver Optimization improves maneuverability in unsteady flight; Damage Tolerant Control re-allocates control forces and moments to allow for a greater chance of survival in case of damage to the aircraft. This paper describes the implementation and testing of these algorithms in piloted simulation of an operational mission using the X-49 winged compound aircraft. The results show that, with ADAPT™ on, the operational mission can be flown faster while burning less fuel, resulting in increased aircraft endurance and range. Maneuvering flight with ADAPT™ can be flown faster as well. In the case of simulated aircraft damage, aircraft without ADAPT™ were unable to recover and crashed in every case tested, while those with ADAPT™ were able to recover, fly to an emergency airfield, and land.

INTRODUCTION

Modern aircraft fly faster, further, and with higher levels of efficiency, reliability, and survivability (Ref. 1). However, in recent years, hundreds of rotorcraft and lives have been lost during military combat and non-combat operations (Ref. 2), meaning that further improvements in safety and survivability are still necessary. ADAPT™ (Adaptive Digital Automated Pilotage Technology) is a flight control software package designed to improve the safety, survivability, performance, and affordability of advanced Vertical Takeoff and Landing (VTOL) capable aircraft. These improvements are driven by automatically re-allocating forces and moments on the aircraft produced by control effectors. Since many advanced VTOL aircraft have additional control effectors compared to a conventional rotorcraft, finding the optimal position of each control effector is time-consuming for a human pilot, resulting in a large amount of pilot workload. By using a digital flight control system with ADAPT™ to find the optimal positions for each control effector (based on given objectives), the aircraft can fly with improved performance while reducing the workload for the human pilot. Alternatively, if the aircraft sustains damage to its control effectors, some combination of still-operational effectors may be able to allow the pilot to retain control and continue flying or land safely.

ADAPT™ DESCRIPTION

The ADAPT™ flight control package (shown in schematic form in Fig. 1) currently includes three main flight control modes: Fly-to-Optimal (FTO), Vehicle Maneuver Optimization (VMO), and Damage Tolerant Control (DTC). In addition, an Executive Manager acts as supervisory control system to manage these flight control modes based on pilot commands, aircraft states, and data from external sensors. A Configuration Manager component allows for the ADAPT™ software to operate on aircraft of various aircraft and control effector configurations.

Fly-to-Optimal (FTO)

Fly-to-Optimal is an algorithm designed to select one trim state out of the many possible using the multiple control effectors on an advanced configuration VTOL aircraft, as described in Refs. 3-6; the selected trim state is chosen based on performance objectives specified by the pilot. For the results presented below, five FTO modes with different

Presented at the VFS Modern Ops. with Future Integrated VMS Technical Meeting, West Chester, PA, Aug. 19-20, 2021.

performance objectives were demonstrated; these are listed in Table 1 below.

Trim optimization can be achieved because advanced VTOL aircraft with multiple control effectors (such as the compound rotorcraft used in this study) have a range of control options to balance forces and moments on the aircraft. Using the example of the compound rotorcraft flown in this study, the sharing of lift and propulsive forces between the main rotor, lifting wing, and propeller can be varied continuously over a range of trim conditions, resulting in a significant change in aircraft performance. Pitching moment balance between the main rotor and horizontal tail surfaces (including elevators) can also affect performance as well as structural loads on the main rotor hub. FTO seeks to find the optimal trim automatically during long-duration trimmed flight. Without this autonomous feature, trim optimization would be a high workload task for the pilot.

Vehicle Maneuver Optimization (VMO)

Vehicle Maneuver Optimization (VMO) is designed to increase aircraft maneuverability by commanding all available control effectors in an optimal manner (Ref. 7). When engaged, VMO uses aircraft state data to anticipate the type of maneuver being performed (turn, pull-up or push-over, climb / descent, or acceleration / deceleration); based on the maneuver type, VMO then uses pre-computed tables to select the optimum control effector positions which would result in an optimal maneuver. For the compound rotorcraft in this study, VMO optimizes deployment of wing flaperons and elevators to minimize power in turns and pull-up maneuvers. The deployment schedules are based on anticipated normal load factor, which is computed from the pitch rate command in the primary flight control system.

In addition, when VMO is engaged, the collective lever acts as a power command inceptor that controls both main rotor collective and propeller thrust. This serves two purposes: the pilot is relieved from manually controlling propeller thrust, and power required is limited, which inherently avoids limit exceedances related to engine torque, power, and temperature.

Damage Tolerant Control (DTC)

Damage Tolerant Control is designed to alter the nominal allocation of the aircraft flight controls to mitigate the effects of damage to the aircraft, whether that damage is the result of enemy fire, or mechanical or other failures. Aircraft damage is tolerated by re-allocating the forces and moments generated by the aircraft's control effectors used in nominal operation to an alternate method that uses the control effectors available in the aircraft's damaged state, whether that differs from nominal operation with control effectors that have reduced effectiveness or are missing altogether.

The current methods that ADAPT™ DTC uses to tolerate damage include pseudoinverse (PI) control allocation / re-allocation, which is a deterministic method that distributes control among multiple redundant control effectors, as well Neural Network Aircraft Control (NNAC), which uses an adaptive neural network (ANN) algorithm that adjusts feedback compensation to enhance command model tracking; both methods are outlined in Ref. 8.

For the ADAPT™ system to engage DTC for control relocation quickly after damage is incurred, reliable methods to detect and identify the type and severity of damage are necessary. In the case of frozen servos (such as the main rotor servo failure encountered in the simulated operational mission below), a comparison of commanded and actual servo positions over a short time was used. For other damage cases tested with DTC, including tail damage (with reduced yaw effectiveness) and a severed tail driveshaft, separate detection algorithms are currently under development, as described in Ref. 9.

Executive Manager

The ADAPT™ Executive Manager is designed to contain the logic to select which components of FTO, VMO, and DTC are to be engaged based on aircraft states, pilot inputs, and sensor data. Primary functions of the Executive Manager include:

- Engagement and disengagement of ADAPT™ components (FTO and VMO) based on direct pilot input via instrument panel buttons, indirect pilot input via cyclic and collective position and rate or monitoring of the current aircraft state.
- Engagement of DTC based on receipt of a damage signal from either the damage detection routines described above or from pilot input.
- Selection and deselection of autopilot functions based on the currently engaged ADAPT™ component.
- Transfer of information on ADAPT™ components for use in the instrument panel displays.

Configuration Manager

The ADAPT™ Configuration Manager is designed to inform the ADAPT™ software suite of the current aircraft configuration, including available control effectors and their properties (axes effected, and lift, drag, thrust, and moment characteristics), properties related to the existing aircraft flight control system and autopilot, information on available external sensor inputs, and other aircraft properties relevant to operation of ADAPT™.

SIMULATION ENVIRONMENT

Aircraft Simulation Model

The X-49S (where S stands for simulated) six degree-of-freedom model used for this simulation is a highly modified H-60 Genhel-PSU model (Ref. 10). The X-49S model integrates actual X-49A flight test data and wind tunnel data into a UH-60A model and thus, accurately represents flight characteristics and performance of the actual X-49A aircraft (Ref. 11). In the X-49S simulation model, the empennage, tail rotor and stabilator modules of the UH-60 are disabled; and Vectored Thrust Ducted Propeller (VTDP) and wing modules are added (including flaperons on the wings and the rudder/sector and internal/external elevators on the VTDP, as shown in Fig. 2); the VTDP is connected to the drive system module; fly-by-wire (FBW) control laws are implemented; and the mass properties and flat plate drag area are adjusted to accurately reflect the X-49S configuration. As illustrated in Fig. 2, the external elevators on the VTDP are a feature of the X-49S that is not present on the actual X-49A aircraft; these elevators were added to allow for greater pitch authority at the tail. Interactions between modules include rotor downwash corrections to the VTDP and wing modules and wing downwash corrections for the VTDP module.

For accurate simulation of real-world conditions that come into play during an operational mission, two specific features were added to the aircraft simulation model. The model allows for variable weight, including the addition or removal of passengers or cargo (as included as a part of the operational mission), as well as fuel burn, (calculated by integrating the fuel flow values produced by the engine module). In addition, the model allows for varying atmospheric conditions; in this study, this variation was limited to changes in density altitude corresponding to changes in true altitude for consistency between flights.

To evaluate the effectiveness of Damage Tolerant Control, it is necessary to model damage and related effects on the aircraft. Various damage cases have been included in the X-49S simulation; for the tests described below, the damage case is a frozen main rotor lateral actuator. For the servo damage cases, there are two main parameters: one is damage percentage, which corresponds to the reduction of servo travel speed (in the results below, the damage percentage is always set to 100% to define that the lateral servo is frozen in place); the other parameter is the location where the servo is frozen in place. The location where the servo is frozen is important in the case of a frozen main rotor lateral servo because the attachment point to the swashplate becomes a pivot point that restricts the motion of the swashplate; even though the forward and aft servos can move normally, the swashplate is constrained by being pinned at the lateral servo connection point with possible restrictions in the movement range of the operational forward and aft servos. For the results below, the lateral servo was always frozen in the same position within its travel range for the sake of repeatability.

Flight Simulation Facility

All simulation flights presented below utilized the full-motion simulator located at the Penn State Vertical Lift Research Center of Excellence. Motion was not required and not used for these tests. Computers that provide flight dynamics calculations via the Genhel-PSU model (described above), image generation, cockpit displays, and an interface to the control loaders are linked via shared memory, which allows instantaneous data sharing over the local network. Out-the-window images are produced using X-Plane and displayed on a 5-meter diameter spherical screen using a 6-channel projection system. Cockpit displays are shown on dual 23-inch touchscreens to allow for additional cockpit buttons or dials; the touchscreens are mirrored, so that the same image is shown on both screens simultaneously. The control loading system has four fully programmable axes that are integrated with sticks and linkages of the cockpit cab.

OPERATIONAL MISSION DESCRIPTION

The results below detail a simulated operational mission flown by three pilots, once each with ADAPT™ off, and once with ADAPT™ on. Individual mission segments (described in Table 2 below) were used to exercise FTO (all five modes), VMO, as well as DTC after aircraft damage occurs.

The simulated mission was executed by flying along a series of waypoints, as shown on the map in Fig. 3, with the pilots following a given mission profile. Waypoints are displayed to the pilots in the simulator using orange hoops placed in the proper locations and altitudes within the out-the-window display (Fig. 4).

Table 2 lists the individual segments of operational mission along with the ADAPT™ component under test. The initial objective of the mission is to perform a troop insertion; after flying a segment where the pilots were instructed to minimize fuel flow, the aircraft enters a terrain course leading to the insertion zone. Following the troop insertion, the pilots are instructed to perform an emergent aircrew recovery mission; segments of the emergent mission include a maximum airspeed dash to a staging location, a 10-minute loiter at the staging location while the downed aircrew is located, and an additional terrain course between the staging location and the aircrew recovery location. After recovering the downed aircrew, the pilots fly a third terrain course before returning to base at best range airspeed.

Before reaching the base, the aircraft encounters damage in the form of a frozen main rotor actuator, a damage case that could result from either battle damage (i.e., a bullet passing through the actuator) or a mechanical fault (i.e., loss of actuator fluid pressure). Upon the initiation of damage, pilots are instructed to recover the aircraft, fly to the nearest available airfield, if possible (or land where necessary if required), and perform a safe landing, which could include a run-on landing if necessary.

PILOTED SIMULATION RESULTS

The simulated operational mission was flown by two Army experimental test pilots and one Piasecki test pilot, once each with ADAPT™ on and once with ADAPT™ off. Fig. 5 shows a comparison of the two flights by one pilot; the map shows the path flown, including the waypoint markers (orange hoops), and locations where FTO, VMO, and DTC are engaged or disengaged for the flight with ADAPT™ on. Also shown is the path flown after the aircraft incurs damage and flies to the nearest emergency airfield. The black dashed line and related waypoints show the intended route to be flown if the aircraft was not damaged. The time histories show when each ADAPT™ component (either FTO, VMO, or DTC) was enabled during the flight with ADAPT™ on; for the flight with ADAPT™ off, the aircraft is in Standby state, meaning that no ADAPT™ component is active. Each mission segment is labeled in order by letter. The results of each segment are discussed individually below.

Mission Segment A Results – Minimum Fuel Flow (FTO Mode 1)

Relevant performance results for Mission Segment A for each pilot are compiled in Table 3. In Mission Segment A, pilots were instructed to navigate the waypoints at 120 KIAS while minimizing the aircraft's fuel flow (either manually with ADAPT™ off, or by using FTO Mode 1 with ADAPT™ on). As the table shows, the aircraft maintained approximately 120 KIAS over the course for all flights. With ADAPT™ on, the aircraft had a reduced fuel flow compared to the manual flights by each pilot (with the reductions ranging from 6.8% to 17.6% with an average of 10.6%). In addition, the total fuel burn with ADAPT™ on was reduced by an average of 9.5%.

Mission Segment F Results – Reduced Main Rotor Hub Moment (FTO Mode 2)

Relevant performance results for Mission Segment F are compiled for each pilot in Table 4. Pilots were instructed to fly this segment at 120 KIAS. For flights with ADAPT™ on, the pilots were also instructed to engage FTO Mode 2 with the intention of reducing vibration (measured in this study by using main rotor hub pitching moment magnitude as an analog) for assessment and treatment of the recovered aircrew.

As shown in the table, the aircraft maintained approximately 120 KIAS for all flights. For the flights with ADAPT™ on, the magnitude of the main rotor hub pitching moment was reduced by an average of 76.2%, with a range of 60.6% to 91.1%. Although the goal was to reduce the hub moment magnitude, the average fuel burn during this segment was also reduced by an average of 3.0%.

Mission Segment D Results – Best Endurance Airspeed (FTO Mode 3)

Table 5 lists the relevant performance figures for each pilot. Pilots were instructed to fly a 10-minute loiter at best endurance airspeed; for the cases, with ADAPT™ on, best endurance airspeed was to be found using FTO Mode 3. The best endurance airspeed found by the pilots flying manually was approximately 66 KTAS, while that found by FTO Mode 3 was approximately 68-70 KTAS; the true best endurance airspeed may be slightly different for each case due to varying altitudes and aircraft weights. The average fuel flow for the ADAPT™ on flights was found to be 15.9% less than the flights with ADAPT™ off (with a range of 11.1% to 21.5% reduction in fuel flow). Total fuel burn was also reduced by an average of 23.4% with ADAPT™ on, however, one pilot flew the ADAPT™-on loiter pattern for less than 10 minutes, which in part explains the larger total fuel burn reduction of 27.4%. As a related result of the reduced fuel flow, specific range for this segment was 21.5% higher on average with ADAPT™ versus ADAPT™ off.

Figure 6 illustrates the time histories for one pilot's flights. With ADAPT™ off, the pilot slows without adjusting flap or elevator position or rotor speed. With FTO Mode 3 engaged, ADAPT™ slows the aircraft to an initial estimate for best endurance trim by reducing airspeed and adjusting the symmetric flap and elevator positions. The FTO algorithm then searches on main rotor speed, then symmetric elevator, and finally symmetric flap position.

Mission Segment H Results – Best Range Airspeed (FTO Mode 4)

Relevant performance details for the best range segment of the operational mission are compiled for each pilot in Table 6. Although this segment for each flight ended with the aircraft incurring damage, the location where the damage occurred was kept consistent between flights (2 nautical miles before Waypoint 74), so the time and distance flown remained approximately constant.

For all flights, either with ADAPT™ on or flown manually by the pilot, the best range airspeed was found to be approximately 123 KTAS, on average. With ADAPT™ (FTO Mode 4) on, however, average fuel flow was reduced by 7.0% and total fuel burn reduced by 9.6%, resulting in an average increase of specific range by 10.6%.

Mission Segment C Results – Maximum Level Flight Airspeed (FTO Mode 5)

Table 7 lists the results of the performance details for the maximum airspeed (V_H) dash segment of the operational mission for each pilot. On average, the flights with ADAPT™ on reached 168.6 KTAS, an increase of 2.7% over the manually piloted flights. One pilot was able to fly faster manually than with ADAPT™ on (166.6 KTAS versus 163.1

KTAS); this is possible since ADAPT™ reserves a small percentage of total power to avoid exceeding limits including power, torque, and turbine gas temperature. Also, the FTO algorithm may abandon a trim search on a particular control effector if limits are encountered, which could result in a sub-optimal trim state with a lower maximum airspeed (note that the ADAPT™ on runs for the other two pilots were both above 171 KTAS).

Mission Segments B, E, and G Results – (VMO)

Table 8 lists relevant performance figures for the combined total of all three terrain flights segments (B, E, and G) for the operational mission. Pilots were instructed to fly the terrain courses at maximum safe airspeed; for the ADAPT™ (VMO) on flights, the average airspeed was 6.2% higher than with ADAPT™ off (122.9 KTAS versus 115.7 KTAS). A side benefit of VMO resulted in reduced average fuel flow (1.9% less with VMO than without) and total fuel burn (7.9% less with VMO than without) for these sections; combined with the increased airspeed, the reduced fuel flow resulted in the ADAPT™ on flights having an 8.2% increase in specific range as well.

Figure 7 shows a map and time histories of one of the three terrain course segments (Segment E) for one pilot with ADAPT™ on and off. The time histories show that the pilot did not adjust flap or elevator position when flying manually, while ADAPT™ (VMO) automatically adjusted flap and elevator. The larger flap and elevator movements at the end of the time traces in both cases are due to the aircraft autopilot dropping the flaps and adjusting the elevators during deceleration to landing to recover the downed aircrew after Segment E.

Mission Segment J Results – Damage Detection and Tolerance (DTC)

Relevant performance parameters are listed Table 9 for Segment J of the operational mission for each flight after the aircraft is damaged. As shown, with ADAPT™ off, the aircraft only flies for approximately 12 seconds after damage, since none of the three pilots were able to regain control after damage before crashing. With ADAPT™ (DTC) on, all three pilots were able to recover the aircraft, fly to the nearest emergency airfield (as shown in Fig. 3, 5, and 8), and perform a rolling landing. A rolling landing was conducted in each case since decreasing forward airspeed to come to a hover reduces the effectiveness of the additional aerodynamic control effectors (flaperons and elevators).

Figure 8 shows a map layout and time histories for the damage segment (Segment J) for one pilot with and without ADAPT™ (DTC). The map shows that without DTC, the pilot was not able to recover and move towards the emergency airfield, while with DTC, the pilot was able to recover and land at the emergency airfield. The time histories show that the differential flaperon and elevator is used after damage as a part of the control re-allocation away from the main rotor servos, including the fixed lateral servo and reduced travel forward and aft servos.

In order to evaluate the survivability of a landing after damage, the Sullivan Survivability Scale (Fig. 9) as described in Ref. 8, is used. This scale is based on the Cooper-Harper handling qualities rating scale; by asking the pilot a series of questions and applying a set of quantitative metrics, a survivability score on a scale of 1 to 5 (with 1 being the best and 5 being the worst) is assigned. For the flights above, all three flights without ADAPT™ are rated as a 5, since each pilot was unable to recover before the aircraft crashed. All three flights with DTC active were rated as a 3; although all three flights laded with a rate of descent under the 1,000 foot per minute threshold, the airspeed at landing for all three flights was faster than the 60-knot limit specified for the UH-60, which also applies to the X-49S (Ref. 12).

Overall Mission (Segments A-H) Results

Table 10 compiles the relevant performance metrics for all three pilots during Segments A through H of the operational mission (Segment J after damage is not included for fair comparison between flights with and without DTC). Even though each segment of the operational mission has a different objective, the flights with ADAPT™ resulted in reduced total fuel burn (836.6 lb versus 949.4 lb for a reduction of 11.9%), reduced average fuel flow (848.0 lb/hr versus 944.3 lb/hr for a reduction of 10.2%), and increase specific range (0.1301 nm/lb of fuel versus 0.1149 nm/lb of fuel for an increase of 13.2%), as well as a slightly faster average airspeed (110.3 KTAS versus 108.5 KTAS, for an increase of 1.7%),.

CONCLUSIONS

Based on the results of a simulated operational mission flown with and without ADAPT™ by three pilots, the following conclusions can be drawn, for an advanced VTOL aircraft with multiple control effectors (in this case, a winged compound helicopter):

1. Control algorithms (called Fly-to-Optimal within ADAPT™) have been developed to perform real-time trim optimization without constant input from the pilot. Different optimal trim states can be found depending on the desired performance objective. The FTO algorithms led to overall improvements of more than 10% in average fuel flow, total fuel burn, and specific range.
2. Another set of control algorithms (called Vehicle Maneuver Optimization within ADAPT™) has been developed for optimal maneuver performance of an aircraft, again without constant input from the pilot. The VMO algorithm led to the ability to maneuver through terrain courses at an average airspeed 6% faster than without VMO.

3. If an aircraft's control effectors are damaged or destroyed, control forces and moments can be re-allocated to other effectors to allow the aircraft to recover from damage and land with a greater chance of survival.

Author contact: Derek Bridges bridges_do@piasecki.com

ACKNOWLEDGMENTS

The authors would like to thank the U.S. Army Combat Capabilities Development Command Aviation & Missile Center for its support under contract W911W6-17-D-0001, and the test pilots who participated in the simulation study: Craig Ernst, Grey Hagwood, and Carl Ott.

REFERENCES

1. Chase, N. A., "Future Vertical Lift Overview," American Helicopter Society International Specialists' Meeting on Rotorcraft Propulsion, Williamsburg, VA, October 2015.
2. Couch, M. and Lindell, D., "Study on Rotorcraft Safety and Survivability," American Helicopter Society 66th Annual Forum, Phoenix, AZ, May 2010.
3. Ozdemir, G. T., Horn, J. F., and Thorsen, A.T., "In-Flight Multi-Variable Optimization of Redundant Controls on a Compound Rotorcraft," Paper AIAA 2013-5165, AIAA Guidance, Navigation, and Control (GNC) Conference, Boston, MA, August 19-22, 2013.
4. United States Patent Number 7,348,269., "Compound Aircraft Control System and Method," Inventors: Piasecki, F. N., Greenjack, A. S., and Horn, J. F., October 21, 2008.
5. United States Patent Number 8,463,465., "Compound Aircraft Control System and Method," Inventors: Piasecki, F. N., Greenjack, A. S., and Horn, J. F., June 11, 2013.
6. United States Patent Number 10,831,192., "Control System for an Aircraft," Inventor; Piasecki, J. W., November 10, 2020.
7. Thorsen, A. T., Horn, J. F., and Ozdemir, G. T., "Optimizing Control of a Compound Rotorcraft in Quasi-Steady Maneuvers," Fifth Decennial American Helicopter Society Aeromechanics Specialists' Conference, San Francisco, CA, January 22-23, 2014.
8. Knapp, M. E., Ivler, C. M., Horn, J. F., Johnson, E. N., Lopez, M. J. S., Tischler, M. B., Wagster, J. A., and Cheung, K. K., "Development and Simulation of Damage Tolerant Control Laws for a Compound Helicopter," Paper AIAA 2020-1831, AIAA Scitech 2020 Forum, Orlando, FL, January 6-10, 2020.
9. Lewis, J. A., Iyer, V., Johnson, E. N., "Compound Rotorcraft Yaw Control Fault Detection," Vertical Flight Society's 76th Annual Forum, Online, October 5-8, 2021.
10. Howlett, J., "UH-60A Black Hawk Engineering Simulation Program: Volume I – Mathematical Model," NASA CR 166309, December 1981
11. Geiger, B., Piasecki, F. W., Horn, J. F., Lotterio, M., Schifferle, P., "Challenges of Flight Control in a Compound Helicopter," International Powered Lift Conference 2010, Philadelphia, PA, October 5-7, 2010.
12. Headquarters, Department of the Army., "Operator's Manual for UH-60A Helicopter / UH-60L Helicopter / EH-60A Helicopter," October 31, 1996.

FIGURES

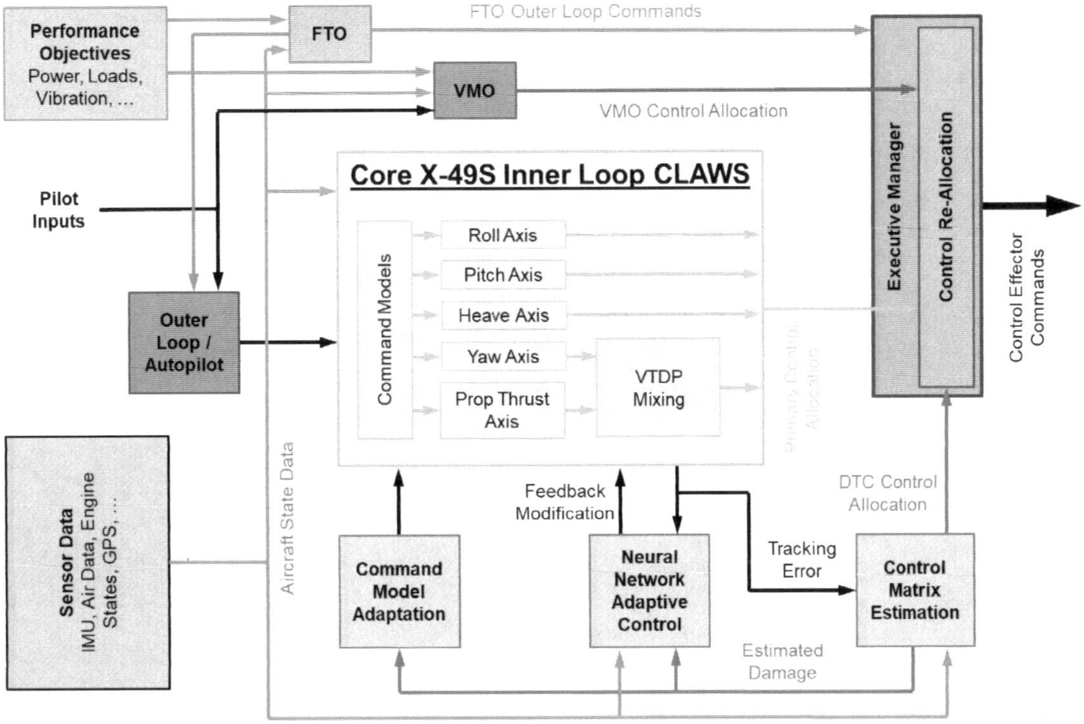

Figure 1. ADAPT™ flight control software schematic diagram.

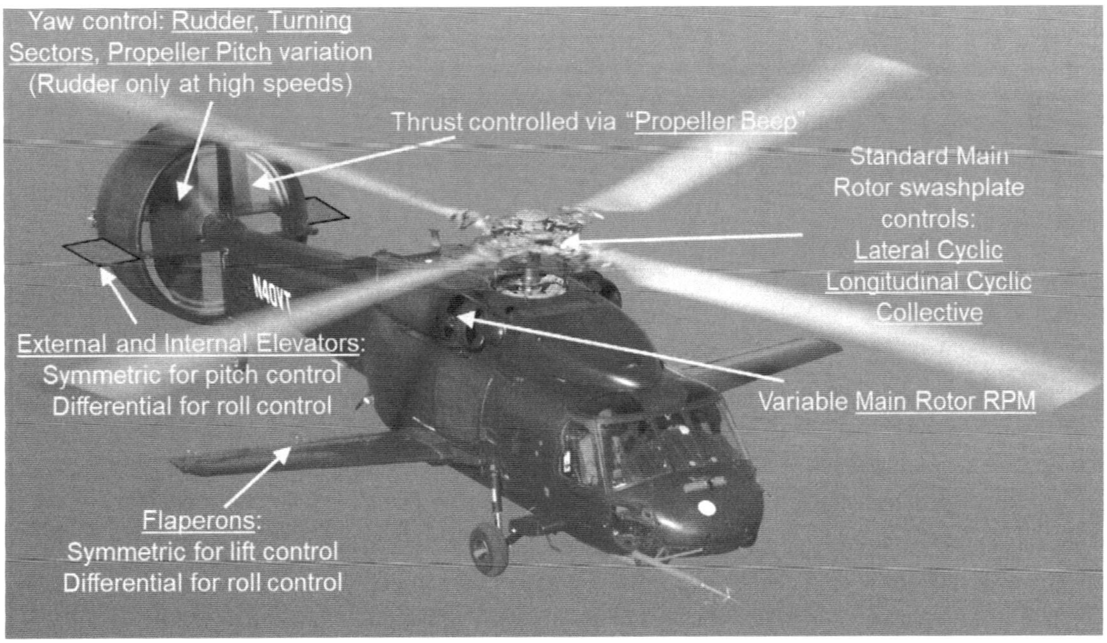

Figure 2. X-49S winged compound aircraft with control effectors labeled.

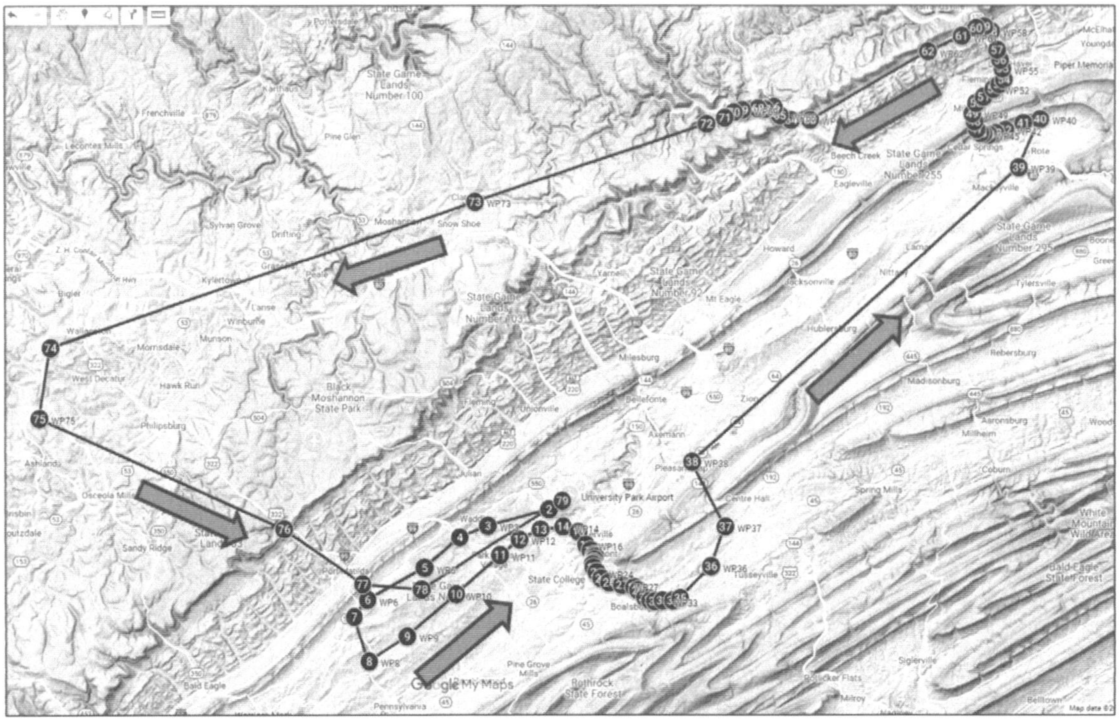

Figure 3. Operational mission map with waypoints and airfields labeled.

Figure 4. Out-the-window view showing hoops used for waypoint navigation.

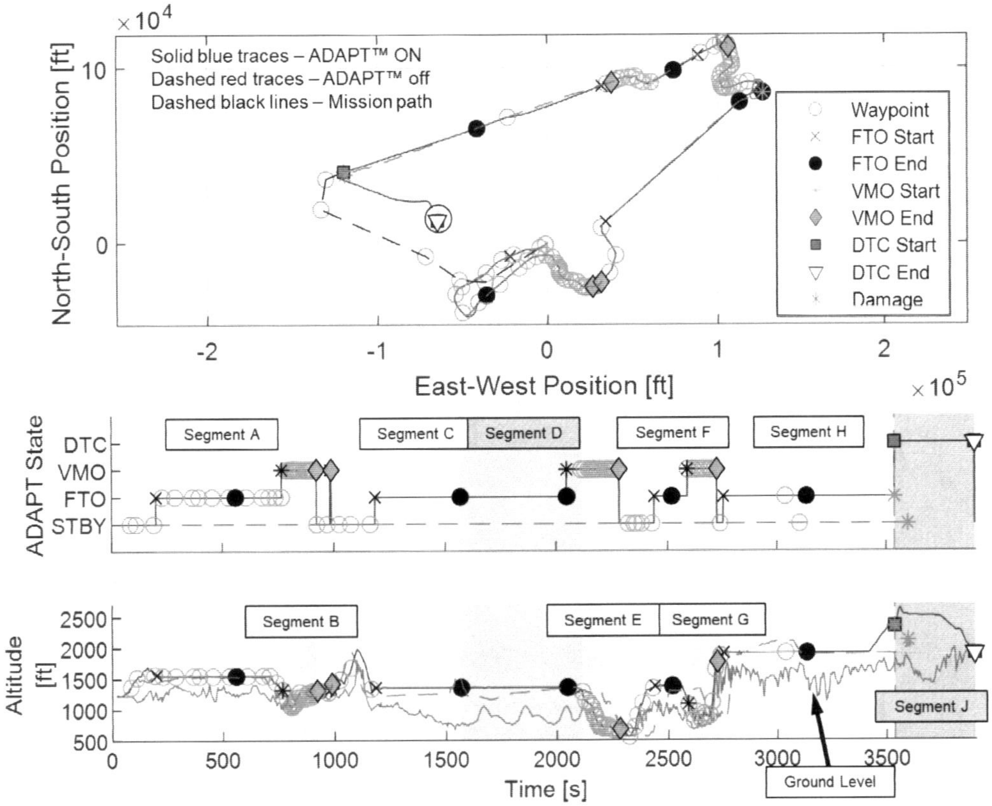

Figure 5. Operational mission map and time history results example.

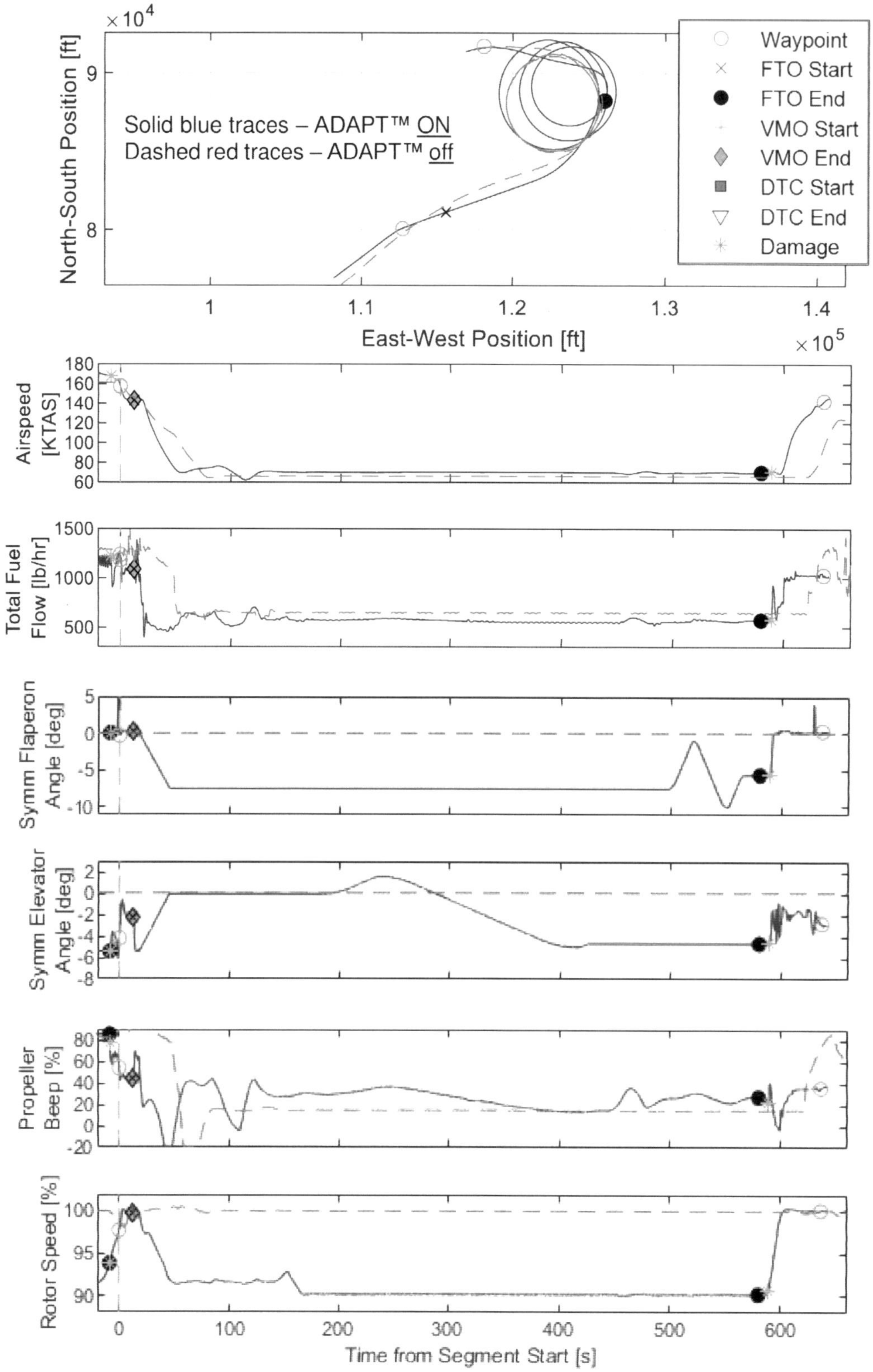

Figure 6. FTO 3 example - 10 minute loiter at best endurance airspeed.

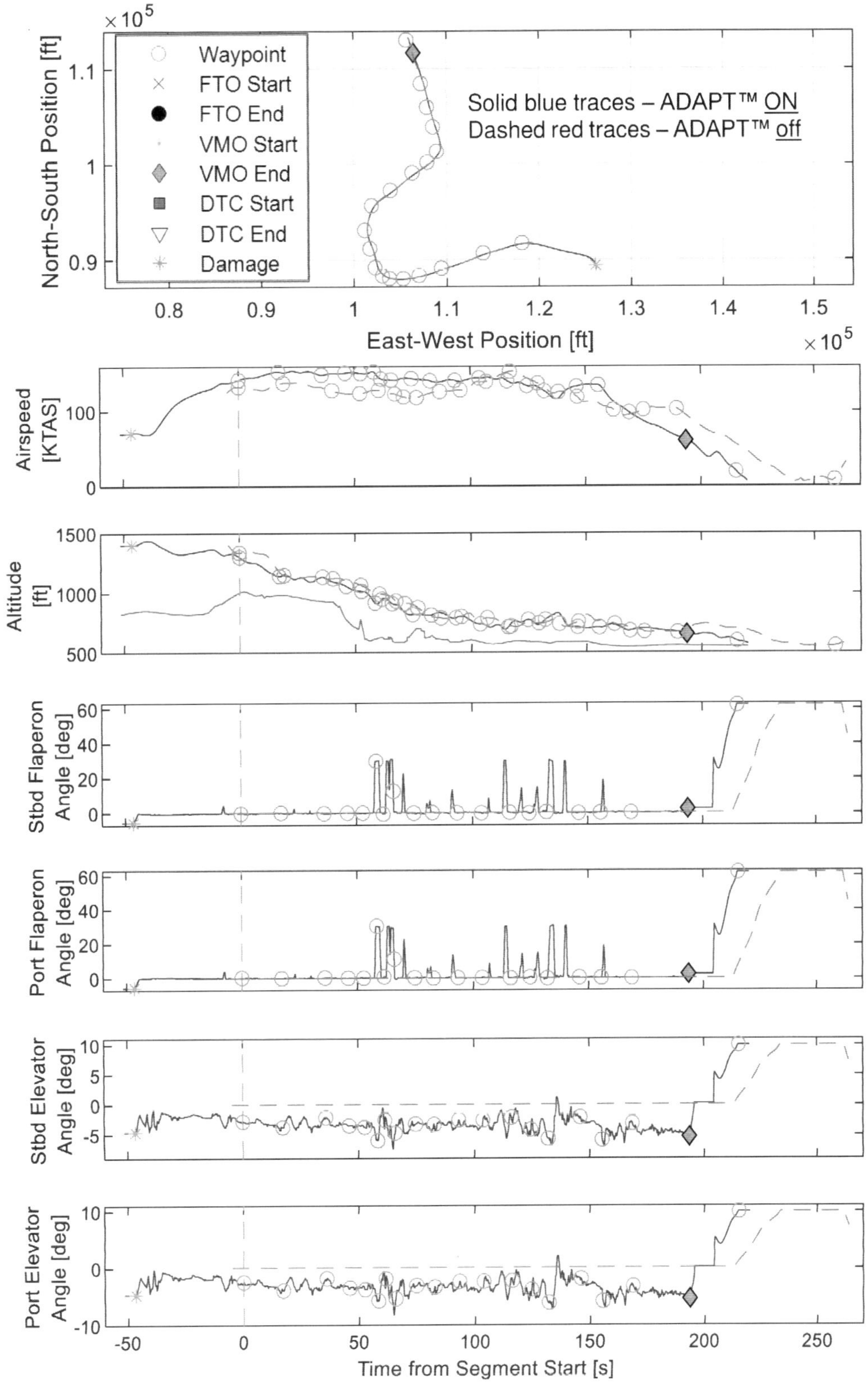

Figure 7. VMO example - Mission segment E terrain course.

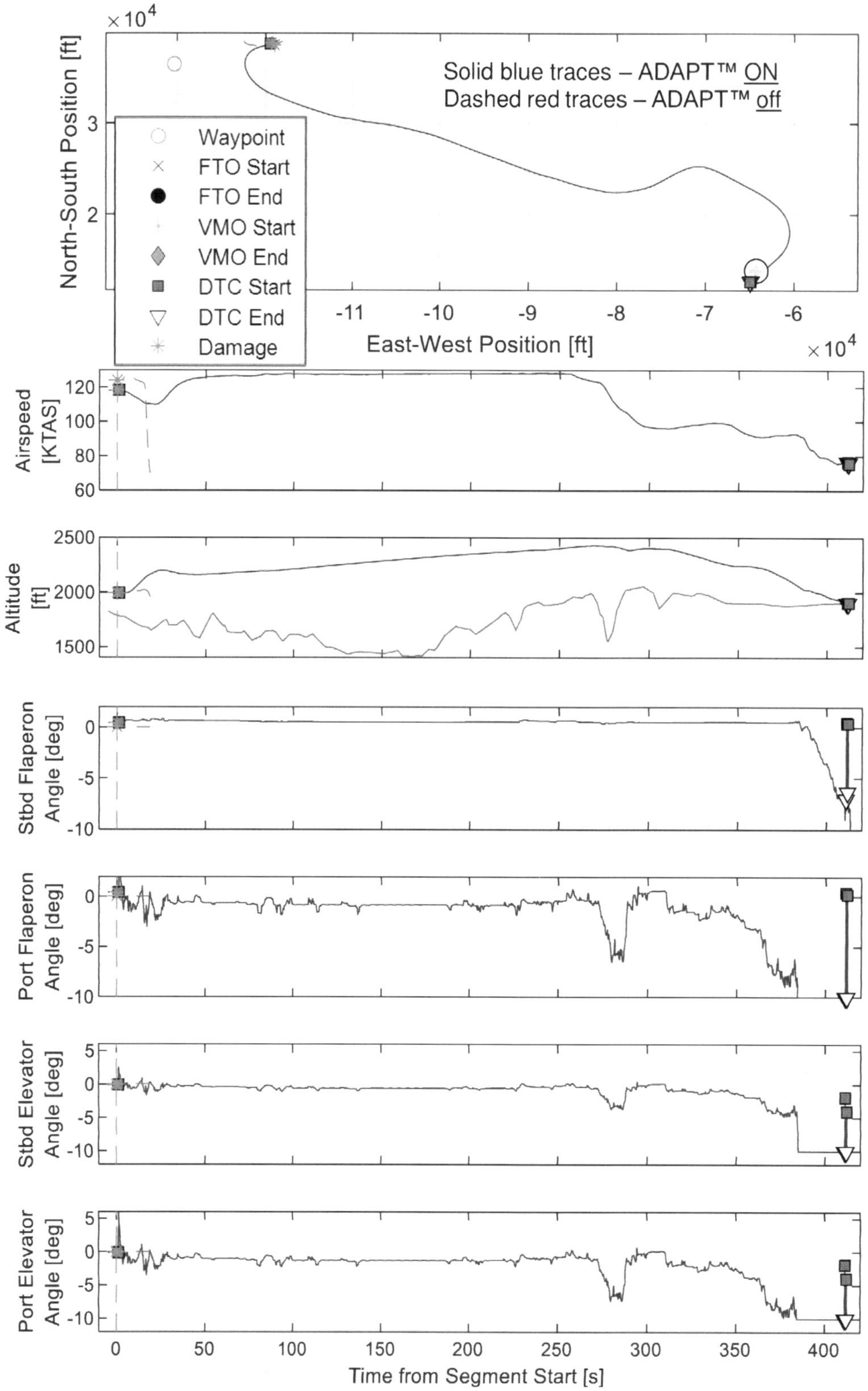

Figure 8. DTC example – Main rotor lateral servo frozen in place.

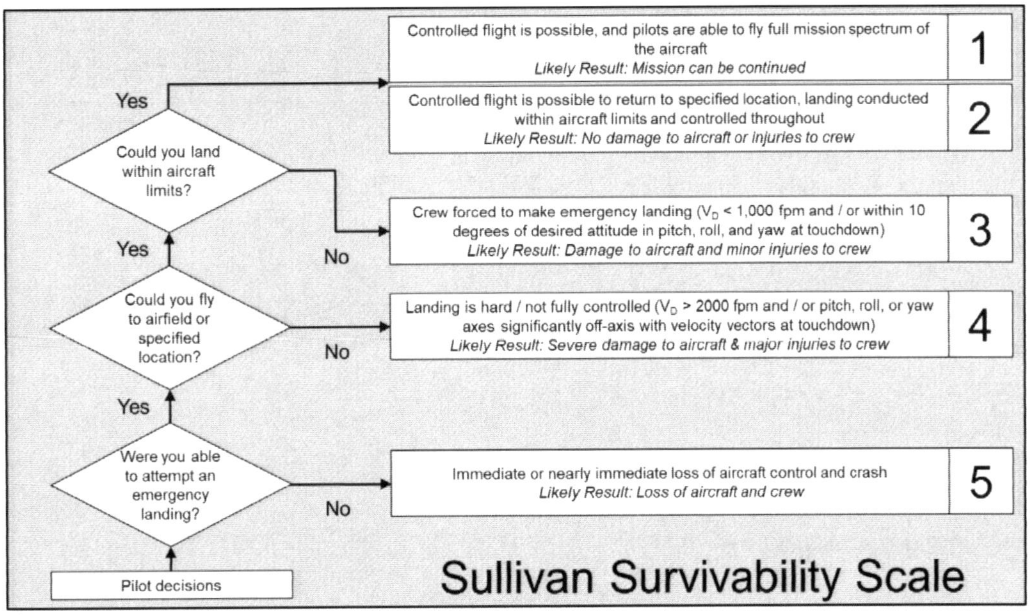

Figure 9. Sullivan Survivability Scale.

TABLES

Table 1. Fly-to-Optimal (FTO) mode descriptions.

FTO Mode Number	Performance Objective	Desired Final Airspeed
1	Minimum fuel flow at speed	Initial airspeed
2	Reduction in main rotor hub moment	Initial airspeed
3	Overall minimum fuel flow	Best endurance airspeed (V_{BE})
4	Overall maximum specific range	Best range airspeed (V_{BR})
5	Maximum level flight airspeed	V_H

Table 2. Operational mission segment descriptions.

Mission Segment	Waypoint Range	Description	ADAPT™ Component Tested
A	WP1–WP14	Low-level course – fly at 120 KIAS	FTO 1
B	WP15–WP34	Terrain course – fly at max. safe speed	VMO
–	WP34–WP37	Land and discharge troops	None
C	WP37–WP39	Overall maximum specific range	FTO 5
D	WP39	Loiter for 10 minutes at best endurance airspeed	FTO 3
E	WP40–WP57	Terrain course – fly at max. safe speed	VMO
–	WP58–WP61	Land and pick up downed aircrew	None
F	WP62–WP63	Fly with reduced vibration / hub moment	FTO 2
G	WP63–WP72	Terrain course – fly at max. safe speed	VMO
H	WP72–WP79	Return to base at best range airspeed	FTO 4
J	Damage location to emergency airfield	Recover and land (if possible) at emergency airfield	DTC

Table 3. Mission Segment A results – minimum fuel flow at speed (FTO Mode 1).

	Pilot 1 ADAPT Off	Pilot 1 ADAPT ON	Pilot 2 ADAPT Off	Pilot 2 ADAPT ON	Pilot 3 ADAPT Off	Pilot 3 ADAPT ON	Average ADAPT Off	Average ADAPT ON
Total Time	9.20 min	9.55 min	9.40 min	9.56 min	9.39 min	9.21 min	9.33 min	9.44 min
Average Airspeed	124.7 KTAS	122.3 KTAS	123.6 KTAS	123.7 KTAS	122.8 KTAS	127.1 KTAS	123.7 KTAS	124.4 KTAS
Total Fuel Burn	155.0 lb	132.6 lb (-14.5%)	165.9 lb	157.3 lb (-5.2%)	164.1 lb	149.0 lb (-9.2%)	161.7 lb	146.3 lb (-9.5%)
Average Fuel Flow	1,010.7 lb/hr	832.7 lb/hr (-17.6%)	1,059.6 lb/hr	987.6 lb/hr (-6.8%)	1,048.6 lb/hr	970.7 lb/hr (-7.4%)	1,039.9 lb/hr	929.9 lb/hr (-10.6%)
Average Specific Range	0.1271 nm/lb	0.1474 nm/lb (+16.0%)	0.1174 nm/lb	0.1258 nm/lb (+7.2%)	0.1172 nm/lb	0.1311 nm/lb (+11.9%)	0.1190 nm/lb	0.1338 nm/lb (+12.4%)

Table 4. Mission Segment F results – reduced main rotor hub moment / vibration (FTO Mode 2).

	Pilot 1 ADAPT Off	Pilot 1 ADAPT ON	Pilot 2 ADAPT Off	Pilot 2 ADAPT ON	Pilot 3 ADAPT Off	Pilot 3 ADAPT ON	Average ADAPT Off	Average ADAPT ON
Total Time	163.86 s	160.05 s	163.21 s	159.72 s	160.45 s	158.05 s	162.51 s	159.27 s
Average Airspeed	121.9 KTAS	124.7 KTAS	122.8 KTAS	125.2 KTAS	124.3 KTAS	126.7 KTAS	123.0 KTAS	125.5 KTAS
Total Fuel Burn	38.5 lb	43.0 lb (+11.7%)	46.9 lb	42.9 lb (-8.5%)	44.0 lb	39.6 lb (-10.0%)	43.1 lb	41.8 lb (-3.0%)
Starting Hub Moment	-5,121.0 ft-lb	+5,127.0 ft-lb	+6,559.0 ft-lb	-1,762.9 ft-lb	+6,286.0 ft-lb	-4,332.4 ft-lb	N/A	N/A
FTO End Hub Moment	N/A	-1,178.4 ft-lb	N/A	-695.1 ft-lb	N/A	-383.9 ft-lb	N/A	N/A
Hub Moment Mag. Delta	N/A	-77.0%	N/A	-60.6%	N/A	-91.1%	N/A	-76.2%

Table 5. Mission Segment D results – minimum fuel flow at best endurance airspeed (FTO Mode 3).

	Pilot 1 ADAPT Off	Pilot 1 ADAPT ON	Pilot 2 ADAPT Off	Pilot 2 ADAPT ON	Pilot 3 ADAPT Off	Pilot 3 ADAPT ON	Average ADAPT Off	Average ADAPT ON
Total Time	11.16 min	9.11 min	10.82 min	10.40 min	11.07 min	10.60 min	11.01 min	10.04 min
Best Endurance Airspeed (V_{BE})	66.3 KTAS	69.6 KTAS	66.4 KTAS	68.4 KTAS	65.9 KTAS	70.2 KTAS	66.2 KTAS	69.4 KTAS
Total Fuel Burn	131.1 lb	95.2 lb (-27.4%)	134.3 lb	101.4 lb (-24.5%)	132.7 lb	108.2 lb (-18.5%)	132.7 lb	101.6 lb (-23.4%)
Average Fuel Flow	705.3 lb/hr	627.2 lb/hr (-11.1%)	744.8 lb/hr	584.9 lb/hr (-21.5%)	719.4 lb/hr	612.5 lb/hr (-14.9%)	723.2 lb/hr	608.2 lb/hr (-15.9%)
Average Specific Range	0.1035 nm/lb	0.1273 nm/lb (+23.0%)	0.1062 nm/lb	0.1263 nm/lb (+18.9%)	0.1031 nm/lb	0.1265 nm/lb (+22.7%)	0.1043 nm/lb	0.1267 nm/lb (+21.5%)

Table 6. Mission Segment H results – maximum specific range at best range airspeed (FTO Mode 4).

	Pilot 1 ADAPT Off	Pilot 1 ADAPT ON	Pilot 2 ADAPT Off	Pilot 2 ADAPT ON	Pilot 3 ADAPT Off	Pilot 3 ADAPT ON	Average ADAPT Off	Average ADAPT ON
Total Time	6.64 min	6.72 min	6.74 min	6.47 min	6.88 min	6.50 min	6.75 min	6.56 min
Best Range Airspeed (V_{BR})	121.4 KTAS	120.5 KTAS	124.4 KTAS	118.4 KTAS	123.5 KTAS	129.7 KTAS	123.0 KTAS	122.9 KTAS
Total Fuel Burn	140.1 lb	138.2 lb (-1.4%)	144.9 lb	125.8 lb (-13.2%)	148.2 lb	127.6 lb (-13.9%)	144.4 lb	130.5 lb (-9.6%)
Average Fuel Flow	1,266.4 lb/hr	1,233.5 lb/hr (-2.6%)	1,289.0 lb/hr	1,166.5 lb/hr (-9.5%)	1,292.8 lb/hr	1,178.3 lb/hr (-8.9%)	1,282.7 lb/hr	1,192.8 lb/hr (-7.0%)
Average Specific Range	0.1280 nm/lb	0.1295 nm/lb (+1.2%)	0.1237 nm/lb	0.1423 nm/lb (+15.1%)	0.1211 nm/lb	0.1408 nm/lb (+16.3%)	0.1243 nm/lb	0.1375 nm/lb (+10.6%)

Table 7. Mission Segment C results - maximum level flight airspeed (FTO Mode 5).

	Pilot 1 ADAPT Off	Pilot 1 ADAPT ON	Pilot 2 ADAPT Off	Pilot 2 ADAPT ON	Pilot 3 ADAPT Off	Pilot 3 ADAPT ON	Average ADAPT Off	Average ADAPT ON
Total Time	6.64 min	6.72 min	6.74 min	6.47 min	6.88 min	6.50 min	6.75 min	6.56 min
Maximum Airspeed (V_H)	166.6 KTAS	163.1 KTAS (-2.1%)	164.2 KTAS	171.1 KTAS (+4.2%)	161.5 KTAS	171.5 KTAS (+6.2%)	164.1 KTAS	168.6 KTAS (+2.7%)
Total Fuel Burn	140.1 lb	138.2 lb (-1.5%)	144.9 lb	125.8 lb (-13.2%)	148.2 lb	127.6 lb (-13.9%)	144.4 lb	130.5 lb (-9.6%)
Average Fuel Flow	1,266.6 lb/hr	1,233.3 lb/hr (-2.6%)	1,289.3 lb/hr	1,166.6 lb/hr (-9.5%)	1,245.1 lb/hr	1,178.5 lb/hr (-5.3%)	1,282.8 lb/hr	1,193.0 lb/hr (-7.0%)
Average Specific Range	0.1315 nm/lb	0.1322 nm/lb (+0.5%)	0.1274 nm/lb	0.1467 nm/lb (+15.1%)	0.1297 nm/lb	0.1455 nm/lb (+12.2%)	0.1279 nm/lb	0.1413 nm/lb (+10.5%)

Table 8. Mission Segments B, E, G combined results – maximum safe airspeed in terrain course (VMO).

	Pilot 1 ADAPT Off	Pilot 1 ADAPT ON	Pilot 2 ADAPT Off	Pilot 2 ADAPT ON	Pilot 3 ADAPT Off	Pilot 3 ADAPT ON	Average ADAPT Off	Average ADAPT ON
Total Time	8.30 min	8.24 min (-0.7%)	9.43 min	8.87 min (-5.9%)	9.74 min	8.56 min (-12.1%)	9.15 min	8.56 min (-6.4%)
Average Airspeed	127.0 KTAS	128.6 KTAS (+1.3%)	112.0 KTAS	116.9 KTAS (+4.4%)	108.2 KTAS	123.2 KTAS (+13.9%)	115.7 KTAS	122.9 KTAS (+6.2%)
Total Fuel Burn	152.7 lb	141.2 lb (-7.5%)	148.4 lb	144.2 lb (-2.8%)	155.2 lb	134.9 lb (-13.1%)	152.1 lb	140.1 lb (-7.9%)
Average Fuel Flow	1,104.4 lb/hr	1,027.8 lb/hr (-6.9%)	944.5 lb/hr	975.7 lb/hr (+3.3%)	956.1 lb/hr	945.5 lb/hr (-1.1%)	1,001.7 lb/hr	983.0 lb/hr (-1.9%)
Average Specific Range	0.1150 nm/lb	0.1251 nm/lb (+8.8%)	0.1186 nm/lb	0.1198 nm/lb (+1.0%)	0.1132 nm/lb	0.1303 nm/lb (+15.1%)	0.1155 nm/lb	0.1250 nm/lb (+8.2%)

Table 9. Mission Segment J results – recovery and landing after damage (DTC).

	Pilot 1 ADAPT Off	Pilot 1 ADAPT ON	Pilot 2 ADAPT Off	Pilot 2 ADAPT ON	Pilot 3 ADAPT Off	Pilot 3 ADAPT ON	Average ADAPT Off	Average ADAPT ON
Total Time	10.85 s	361.70 s	18.50 s	413.60 s	7.70 s	613.95 s	12.35 s	463.08 s
Time (Damage to DTC)	N/A	0.65 s	N/A	1.10 s	N/A	0.65 s	N/A	0.80 s
Airspeed at Landing	N/A	78 KTAS	N/A	77 KTAS	N/A	75 KTAS	N/A	77 KTAS
Descent Rate on Landing	N/A	275 fpm	N/A	335 fpm	N/A	414 fpm	N/A	341 fpm
Sullivan Survivability Scale Rating	5 (crash)	3	5 (crash)	3	5 (crash)	3	5 (crash)	3

Table 10. Overall Mission results (Segments A-H, before damage)

	Pilot 1 ADAPT Off	Pilot 1 ADAPT ON	Pilot 2 ADAPT Off	Pilot 2 ADAPT ON	Pilot 3 ADAPT Off	Pilot 3 ADAPT ON	Average ADAPT Off	Average ADAPT ON
Total Time	59.44 min	58.62 min	60.35 min	60.08 min	61.18 min	58.89 min	60.32 min	59.96 min
Average Airspeed	109.7 KTAS	111.4 KTAS (+1.5%)	109.1 KTAS	108.2 KTAS (-0.8%)	106.7 KTAS	111.3 KTAS (+4.3%)	108.5 KTAS	110.3 KTAS (+1.7%)
Total Fuel Burn	907.3 lb	827.6 lb (-8.8%)	974.2 lb	842.7 lb (-13.5%)	966.6 lb	839.4 lb (-13.2%)	949.4 lb	836.6 lb (-11.9%)
Average Fuel Flow	915.9 lb/hr	847.1 lb/hr (-7.5%)	968.6 lb/hr	841.6 lb/hr (-13.1%)	948.0 lb/hr	855.2 lb/hr (-9.8%)	944.3 lb/hr	848.0 lb/hr (-10.2%)
Average Specific Range	0.1198 nm/lb	0.1315 nm/lb (+9.8%)	0.1126 nm/lb	0.1286 nm/lb (+14.2%)	0.1126 nm/lb	0.1301 nm/lb (+15.5%)	0.1149 nm/lb	0.1301 nm/lb (+13.2%)

Machine-learning Framework for Online Probabilistic Rotor Fault Detection, Identification, and Quantification on a Multicopter using Boom Strain Signals

Airin Dutta PhD Student
Robert Niemiec Lecturer
Fotis Kopsaftopoulos Assistant Professor
Farhan Gandhi Redfern Professor, Director

Center for Mobility with Vertical Lift (MOVE)
Rensselaer Polytechnic Institute, Troy, NY

ABSTRACT

A robust machine-learning (ML) based framework for online rotor fault detection, identification and quantification in multicopters while effectively rejecting the effects of gusts is introduced. The framework employs in-flight time-series data obtained from the a 2 kg gross weight hexacopter flying at a forward speed of 5 m/s. A concise overview of the rational selection of signals from the on-board sensors network, and using those signal properties for training the ML algorithms are presented. This scheme relies on static properties of the local signals, the strain gauge readings at each boom, to distinguish between gust affected healthy aircraft and different rotor fault types. A simple neural network with input as the statistical mean of the above signals can perform simultaneous online rotor fault detection and identification with exceptional decision-making speed and accuracy of fault classification. Next, a simple linear regression predicts the rotor degradation value with 95% confidence intervals from the strain at the boom on which the faulty rotor is mounted. Indicative results at a few test operating conditions (not used in the training phase) demonstrate the generalization capability of the method. It can detect and identify a rotor fault of even 10% degradation while distinguishing it from an aggressive gust of 10 m/s magnitude. The average time of fault detection is less than 0.3 s with classification and quantification accuracy over 97% and 95%, respectively.

INTRODUCTION

Urban air mobility (UAM), enabled by autonomous electric VTOL (eVTOL) aircraft is set to revolutionize urban transport. A NASA sponsored study concluded UAM is a viable option and assessed its available market value at \$500B (Ref. 1). But, the operational success of on-demand aviation service for mass transportation requires absolute safety and reliability. The preliminary safety target of UAM utilizing the airspace over dense urban environments has been established to be twice as safe as driving by Uber Elevate (Ref. 2), with the expected improvement toward airline aviation levels of safety through innovation with full autonomy and large amounts of data from real-world operations after the first generation VTOL aircraft are in production. Therefore, the current interest is towards real-time system-level awareness and safety assurance in UAM aircraft utilizing in-flight data streams. The goal of this line of work is the development of a data-driven and probabilistic rotor fault diagnosis framework in multicopters which will provide online information about rotor faults, critical for control reallocation or vehicle reconfiguration to complete the flight safely.

Multicopters have been identified as a potential platform for future UAM aircraft development due to their rotor redundancy, design flexibility, ability to integrate distributed electric propulsion, and their superior fault robustness and compensation capabilities. These are complex systems that exhibit strong non-linear dynamic coupling between rotors, structural components, fuselage, and control inputs, as well as time-varying and cyclostationary behavior and pose significant system modeling and identification challenges when compared to fixed-wing aircraft. These issues, as well as potential solutions, have been explored in the literature.

There are three main types of rotorcraft faults, classified by their location, namely actuator, sensor, and component faults, which have been addressed by various model representations (linear time-invariant, linear parameter varying, non-linear, and multibody dynamics models and Kalman Filters) (Refs. 3–8), signal processing techniques (in time- and frequency-domain, Kalman filter) (Refs. 9–14), as well as computational intelligence approaches (neural networks, fuzzy logic, hidden Markov models, dynamic Bayesian network, and Gaussian mix-

Presented at the Modern Operations with Future Integrated Vehicle Management Systems, Virtual, August 19–20, 2021 .

ture models, support vector machines) (Refs. 15–24) – for a detailed review the interested reader is referred to (Ref. 25). However, the available studies are either limited by analytical model building with the assumption of physical knowledge of the system available or mostly concentrate on structural faults of blades, propellers, powertrain, etc. in rotorcraft.

Hence, the problem of multicopter rotor fault diagnosis under operating and environmental uncertainty was approached with statistical time series methods.These methods have been used extensively in structural health monitoring (Refs. 26–29) and detection of various types of faults in fixed-wing aircraft systems (Refs. 30,31), due to their simplicity, efficient handling of uncertainties, no requirement of physics-based models, and applicability to different operating conditions. Zhao *et al.* have validated a stochastic hybrid system for health evaluation in multicopters under simulated sensor anomalies (Ref. 32). Previously, Dutta *et al.* achieved fast and accurate online rotor failure detection and identification *via* a novel application of stochastic time-series models and proper statistical decision-making under pre-determined confidence levels to a hexacopter flying forward under different turbulence levels and uncertainty as well as varying forward velocity and gross weight in Refs. 33, 34.

Another class of stochastic methods is the machine learning (ML) techniques, adept in clustering, classifying, and extracting useful features from high-dimensional, noisy data. Neural networks (NN) (Refs. 15–17) and support vector machines (Refs. 23, 24) are being widely used in FDI, to make a decision on the current state of a dynamic system. Ganguli *et al.* (Ref. 16) and Morel *et al.* (Ref. 15) employed NNs to detect and trace faults and defects of helicopter rotor blades using noise-contaminated vibration data. Multicopter rotor structural damage detection and identification and has been demonstrated by Iannace *et al.* in Ref. 17 with acoustic signals and NNs, and by Bondrya *et al.* in Ref. 24 via support vector machines based on measurements of acceleration from the onboard IMU (Inertial Measurement Unit). To address the issues of focus on data-fitting and black-box modeling with ML algorithms, Dutta *et al.* proposed an innovative time-series assisted neural network for online rotor FDI. The features, acting as the input layer of the NN, were extracted from a statistical time-series model of the healthy aircraft for improved physical explainability of the data-driven scheme. Its excellent FDI performance with respect to accuracy and computation time was demonstrated in Ref.35.

To the authors' best of knowledge, papers pertaining to actuator fault quantification is scarce. Fault magnitude information obtained online can facilitate switching over to a more optimal control scheme and planning alternative trajectories with limited control authority depending on the fault severity. Therefore, to fill this research gap the previously developed time-series assisted neural network was extended to determine three discrete fault levels (mild, caution, and urgent) in addition to rotor fault detection and classification while differentiating them from random gusts in Ref.36. To enable continuous fault quantification, a unified statistical time-series framework was developed based on Functionally Pooled (FP) models, that are capable of representing an aircraft dynamics under rotor faults for a continuum of fault magnitudes of a particular rotor fault type (Ref. 37). This framework was applied to a flexible-boom hexacopter, using remote signals from the Inertial Measurement Unit (IMU), and local signals from sensors placed on individual booms. Besides accurate fault detection and identification by either type of signals, improvement in continuous rotor fault quantification was observed for local signals over remote signals.

The objective of this paper is to develop a simple data-driven framework for online rotor fault detection, identification, and quantification in multicopters flying forward under atmospheric turbulence, wind gusts, and uncertainty. A novel application of ML algorithms in the context of probabilistic fault diagnosis on a hexacopter using suitable sensor data (time-series signals) is presented. Its performance with respect to prompt detection, accurate identification, gust distinction capability, and precise quantification under turbulence, random gusts, and rotor faults is investigated.

HEXACOPTER MODEL AND DATA GENERATION

Physics-Based Modeling of Multicopter System

A flight simulation model has been developed for a regular hexacopter (Fig. 1) using summation of forces and moments to calculate aircraft accelerations. This model is used as the source of simulated data under varying operating and environmental conditions, as well as different fault types. Rotor loads are calculated using Blade Element Theory coupled with a 3×4 Peters-He finite state dynamic wake model (Ref. 38). This model allows for the simulation of abrupt rotor failure by ignoring the failed rotor inflow states and setting the output rotor forces and moments to zero.

A feedback controller is implemented on the nonlinear model to stabilize the aircraft altitude and attitudes, as well as track desired trajectories written in terms of the aircraft velocities. This controller is designed at multiple trim points, with gain scheduling between these points to improve performance throughout the flight envelope. The control architecture is detailed in Ref. 39. This control design has been demonstrated to perform well even in the

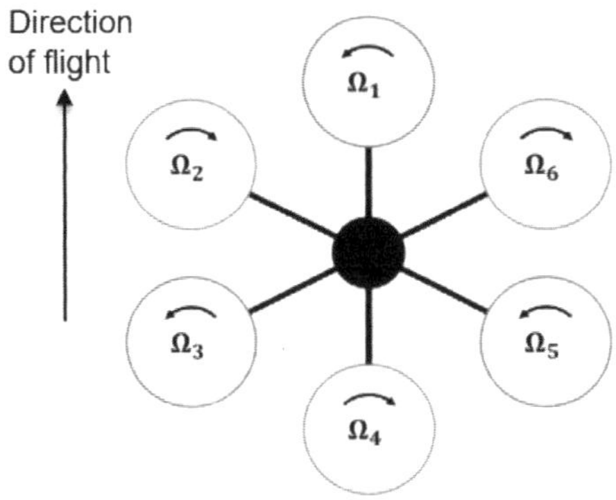

Figure 1: Schematic representation of a regular hexacopter

event of rotor 1, 2 or 6 failure, with no adaptation in the control laws themselves.

The 12 rigid body states are defined in Eq. 1.

$$\mathbf{x} = \{X \quad Y \quad Z \quad \phi \quad \theta \quad \psi \quad u \quad v \quad w \quad p \quad q \quad r\}^T \tag{1}$$

The input vector is comprised of the first four independent multirotor controls for collective, roll, pitch and yaw and is defined in Eq. 2:

$$\mathbf{u} = \{\Omega_0 \quad \Omega_R \quad \Omega_P \quad \Omega_Y\}^T \tag{2}$$

The booms of the hexacopter are modeled as one-dimensional, Euler-Bernoulli beams, with an added tip mass and loading, and are coupled to the rigid body motion of the vehicle. The positive bending deflections in the in-plane and out-of-plane of the hub are illustrated in Fig. 2. Torsion is neglected. The beam equations are discretized in space using the Ritz method with 2 modes in direction, as given in Eq. 3), with polynomial shape functions, ϕ that satisfy the geometric boundary conditions given by Eq. 4. To facilitate the inversion of the mass matrix that arises from this discretization, these polynomials are chosen to be orthogonal. The modes are obtained via eigen-analysis of the beam in a vacuum.

$$\begin{aligned} v &= \sum_{i=1}^{2} \eta_{v_i}(t)\phi_{v_i}(l) \\ w &= \sum_{i=1}^{2} \eta_{w_i}(t)\phi_{w_i}(l) \end{aligned} \tag{3}$$

where, η represents the modal deformations. l is the non-dimensional boom coordinate, representing the distance of a point on the boom from its root normalized by the boom length. It ranges from 0 to 1, denoting the boom root and boom tip, respectively. The geometric boundary conditions are given by:

$$\begin{aligned} w(0) = w'(0) = 0 \\ v(0) = v'(0) = 0 \end{aligned} \tag{4}$$

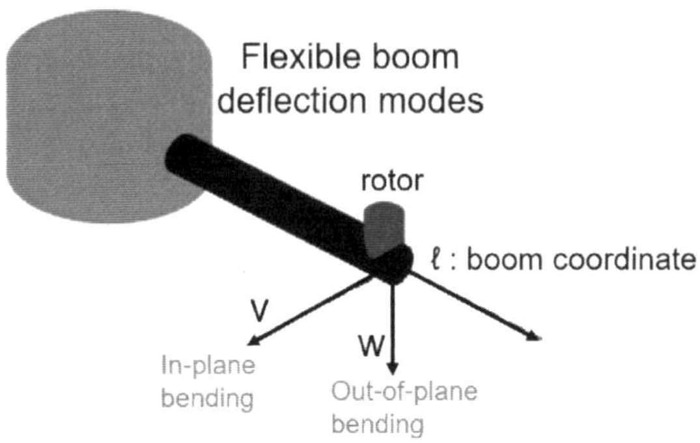

Figure 2: Flexible boom deformation

The 8 flexible states for each boom are defined as follows:

$$\mathbf{x} = \{\eta_{w_1} \quad \eta_{w_2} \quad \eta_{v_1} \quad \eta_{v_2} \quad \dot{\eta}_{w_1} \quad \dot{\eta}_{w_2} \quad \dot{\eta}_{v_1} \quad \dot{\eta}_{v_2}\}^T \tag{5}$$

In the above equations, ' $'$ ' and ' $\dot{}$ ' imply $\frac{d}{dl}$ and $\frac{d}{dt}$, respectively.

The sensor signals can be calculated using the above aircraft states obtained from the flight simulation (See Appendix A).

Incorporation of Turbulence

A continuous Dryden wind turbulence model (Ref. 40) has been incorporated in the flight simulation model to replicate realistic flight conditions. The Dryden model is dependent on altitude, length scale, direction-cosine matrix, body velocity, and turbulence intensity. It outputs the linear and angular velocity components of continuous turbulence as spatially varying stochastic signals. These signals are added to the inflow of each of the rotors by transforming them from the aircraft center-of-gravity to the corresponding rotor locations, to account for the effect of the turbulent wind velocities on the multicopter. The proper combination of the turbulence block parameters determines the fit of the signals to observed turbulence (Ref. 41). In this system, the altitude is taken as 5 m, and the length scale as the hub-to-hub distance of the hexacopter which is 0.6096 m (2 ft). The direction-cosine-matrix and forward velocity are determined from the instantaneous aircraft states during the simulation.

Introduction of Gusts

Next, similar data sets under different magnitude (3- 10 m/s) and directions of gusts for healthy flight under severe levels of turbulence have been generated (Ref. 42). The gusts follow a '1-cosine' shape with gradual increase, followed by a steady-state and gradual decrease

Table 1: Dataset summary

Aircraft state	Number of Datasets			Signal Length
	Turbulence Levels			
Healthy	Severe	Moderate	Light	
	1*+1	1	1	60 s
	Gust Wind Speed (m/s)			
Gusts	8.67	3.46	5, 10	
	8*+8	8	12	20 s
Rotor	Degradation (%)			
1, 2, 6	10,30,...,90	20,40,...,100		
Faults	5*	5		60 s

Sampling frequency: $F_s = 10$ Hz
* indicates training data

(Ref. 43), as shown in Fig.3a. In the simulations, the gusts commence at 4 s with a gradual increase from 0 to its full magnitude in the next 2 s, after which it becomes steady and lasts for 10 s, before gradual decrease to zero again in the next 2 s. Therefore, in the simulated data gusts last from 4-18 s in a span of 20 s of healthy flight. In real flight, the magnitude, direction, time and duration of gusts can be random in nature depending upon the region, altitude, and weather conditions. The magnitude and direction of the gusts considered in this study has been is given in Appendix B, Table 6. The wind velocity directions are denoted by unit vectors with the positive X, Y, Z directions shown in Fig.3b.

Data Generation

The data sets are generated through a series of simulations for different fault types and magnitude under severe of turbulence for the aircraft flying at 5 m/s and having a gross weight of 2 kg. Note that the rotor failures addressed in this paper are: front rotor (rotor 1) and the right-side rotor (rotor 2), and left-side rotor (rotor 6) faults (See Fig. 1). Rotor degradation has been replicated by reducing the commanded speed of that particular rotor by a multiplier ranging from 0 to 1, with '0' denoting 0% rotor degradation , i.e. healthy aircraft and '1' implies 100% rotor degradation or complete rotor failure.

METHODOLOGY

Influence of Gusts on previously developed Statistical Time-series framework for Rotor Fault Diagnosis

In a previous study, a sensor signals based statistical time series method that is capable of effective fault detection, identification, and magnitude estimation within a unified framework for a hexacopter flying under turbulence and uncertainty is presented (Ref. 37). The method is based on the functionally pooled (FP) models (whose parameters are functions of some operating conditions) and proper statistical decision-making schemes. FP models are capable of accurately representing the aircraft dynamics under rotor faults for a continuum of fault magnitudes of a particular rotor fault type. These models for different rotor faults on a multicopter can be identified from available time-series sensor data in the baseline phase. In its inspection phase, there are three distinct steps taking place within a probabilistic framework: step I involves fault detection, step II involves fault identification after fault compensation by the controller, and step III involves precise fault quantification within the identified type of fault. Fault magnitude is continuous, involving an infinite number of potential fault magnitudes ranging from healthy to complete failure. The validity and effectiveness of the method have been assessed via a proof-of-concept application to rotor fault diagnosis with remote and local signals, i.e., z-acceleration signals from the system Inertial Measurement Unit (IMU) and the z-accelerations at the respective boom-tips, respectively. It has been demonstrated that effective fault detection, fault mode identification, and fault magnitude estimation with uncertainty bounds is possible even for "unmodeled" (not used in training) rotor degradation with very limited number of sensor signals (even with a single response signal from the IMU) through powerful signal analysis techniques. An important observation was that fault quantification performance, i.e., the accuracy of fault estimation and tighter uncertainty bounds, was improved via the use of local sensors (boom accelerations) compared to remote sensors (body acceleration).

However, when gusts are introduced in healthy flight, the above method falsely detects the response of the healthy aircraft to gusts as rotor faults, with both the remote and the local signals. To account for random gusts the method needs to be significantly extended, such as identifying a number of time-varying models with the non-stationary signals under gusts of different wind velocity and direction. Moreover, since these methods involve identification of dynamic models from signals, the signals need to be sampled at relatively higher rate after proper filtering to capture the aircraft dynamics and flexible boom natural frequencies.

The current work aims to address the issue of gusts in a simpler fashion, than needed by statistical decision-making with properly identified stochastic models. Previously, the problem of distinguishing gusts from rotor faults had been tackled by using the correlation between aircraft control and response signals in a rigid boom hexacopter, via a systematic analysis detailed in Ref. (Ref. 36). Currently, with the consideration of boom flexibility, local signals contain information about the actual rotor speeds, which can be utilized for rotor fault

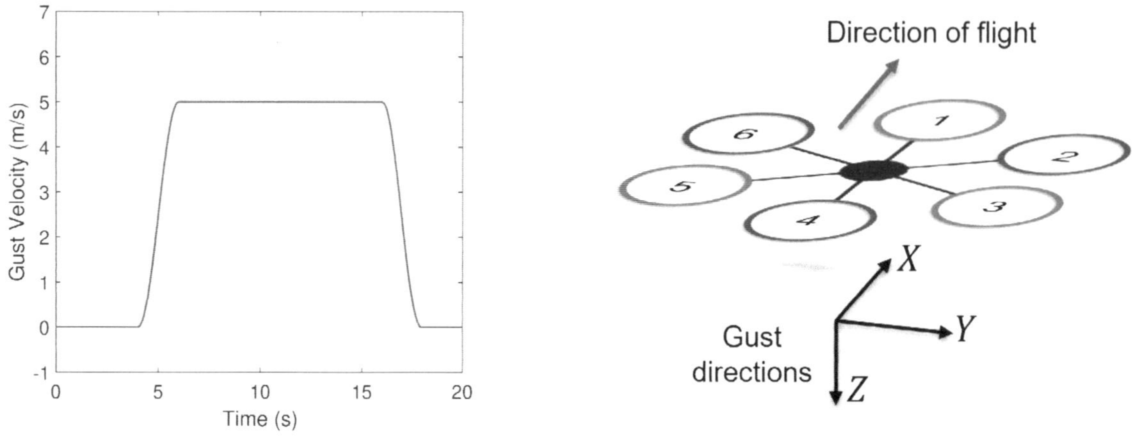

(a) (b)

Figure 4: Gust profile (a) magnitude of gusts with time (b) direction of gust

diagnosis while rejecting gusts. Also, static modeling signals will be attempted to forego the need for higher sampling rate to capture the structural dynamics of the booms.

Selection of Signals

Flight simulation for the hexacopter was performed at operating conditions specified in Table 1. Figures 5a through 7b show time histories of strain gauges placed at the boom roots (where the boom connects to the hub) for cases of healthy flight with and without gusts, and different rotor fault scenarios. The x-axis shows the time in seconds, as the y-axis gives the value of out-of-plane (perpendicular to the rotor hub plane) strain at each of the 6 booms. These local signals, sampled at 10 Hz, will be utilized for the development of the machine-learning based framework of probabilistic fault detection, identification and quantification for their certain characteristics, as discussed in the following paragraphs.

Figure 5a show the strain at each of the boom root in the out-of-plane directions while flying at 5 m/s with gross weight 2 kg under severe turbulence in healthy condition. Though, the signals are stochastic in nature, they possess definite statistical mean value for each of the boom. These mean values of strain on each boom depend on the corresponding the rotor thrust and hence the rotor speeds. Thus, these signals give an insight into the commanded rotor speeds, when there are no rotor faults in the system.

Figures 5b and 6a show the time history of the strain signals under two indicative gusts considered in this study. The gusts commence at 4s and last until 18 s, shown by the black dashed vertical lines. There are no rotor faults in the system. In Fig. 5b, the gusts are coming from all the directions, and result in significant change in the mean values of strain signals, mostly during the steady gust phase lasting from 6 s to 16 s. It can be observed that the absolute value of strains in booms 4 and 6 decreases during gust, mimicking a rotor fault, where the rotor loses thrust. The aim of this study is to prevent such situations being misclassified as rotor faults by considering the response of all the rotors, as reflected in their corresponding strain values. Figure 6a shows a gust 10 m/s coming directly from the direction of flight, which effectively increases the flight speed. This is countered by the controller decreasing the nose-down pitch to maintain flight at the commanded speed of 5 m/s, by increasing the speed of rotor 1 and decreasing that of rotor 6, indicated by the change in mean value of strain.

Figures 6b to 7b shown the strain signals under the considered rotor faults and their varying magnitudes. For the simulation results presented, the rotor faults occurs at $t = 10$ s, as indicated by the vertical black dashed line. It should be noted that due to faults, the signals have a transient response before reaching controller-compensated steady state. In Fig.6b, 30 % degradation of rotor 1 has resulted in the decrease of absolute value of strain on boom 1 due to loss of thrust. To compensate for this fault, rotor 4 also slows down. Similarly, in Figs.7a and 7b, the response of other rotors to rotor 2 and 6 faults, respectively can be observed.

The pattern of how the rotors respond to gusts and faults will be utilized to distinguish them. Also, note that contrary to faults, the change in signals due to gusts last only for its duration. Rotor faults cannot be reversed and hence the change in the signals continue through the entire flight time.

Finally, it can be observed that the rotor fault magnitude has a direct relation with the value of strain on the boom it is mounted upon, as shown in Fig.8. Under complete fail-

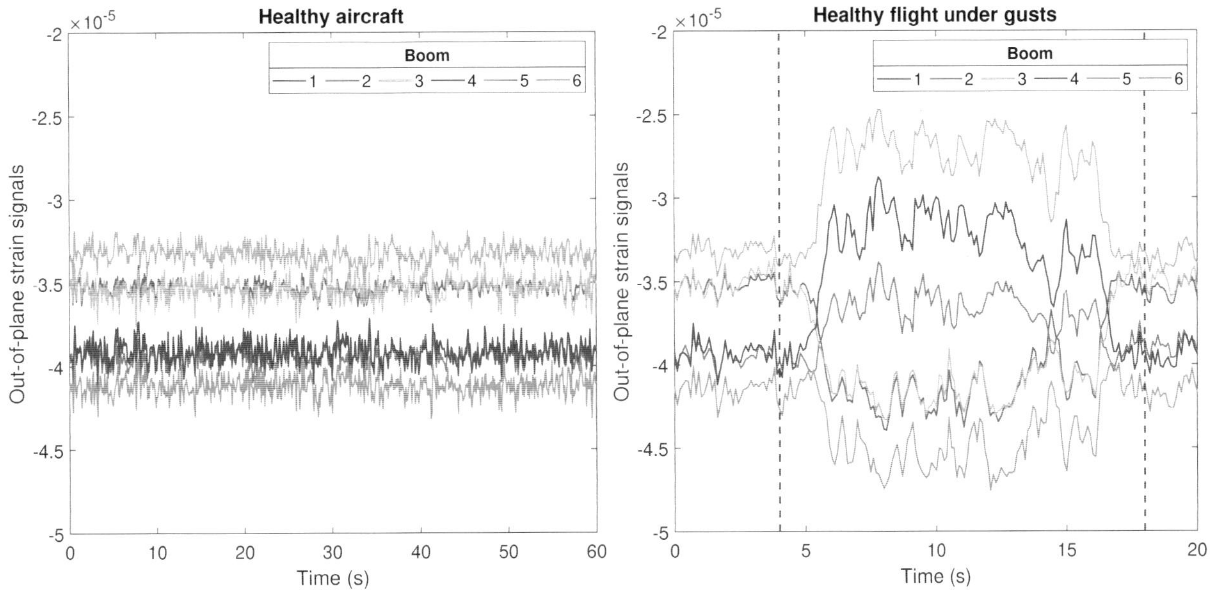

(a) (b)

Figure 5: Time history of out-of- plane strain at each of the booms root under (a) healthy flight (b) healthy flight with gusts, $8.67(-0.57\hat{i}+0.57\hat{j}+0.57\hat{k})$ m/s

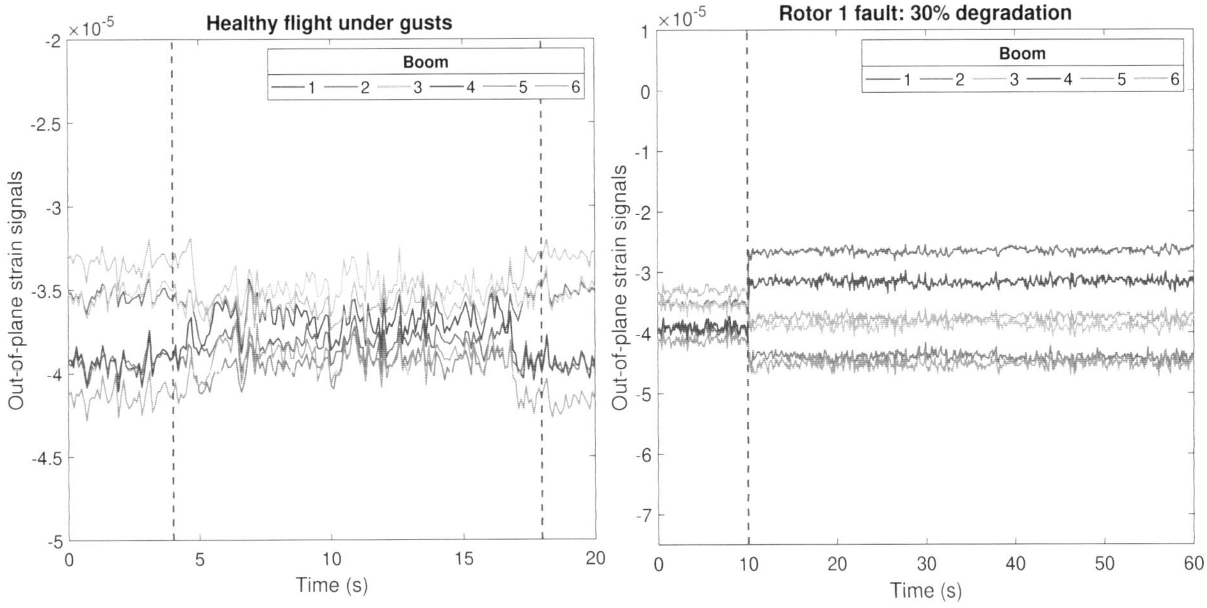

(a) (b)

Figure 6: Time history of out-of- plane strain at each of the booms root under (a) healthy flight with gusts, $10\,(-1\hat{i}+0\hat{j}+0\hat{k})$ m/s (b) 30 % degradation of rotor 1

ure of rotors or 100% degradation, the strain value goes to zero, due to zero thrust generated. The variation of strain on the respective boom can be employed to estimate the fault magnitude, after it is detected and identified.

Development of the Machine-Learning Framework

In this section, the development of a data-driven framework for probabilistic rotor fault diagnosis utilizing the carefully selected in-flight data streams via the application of common ML algorithms is presented. It has been

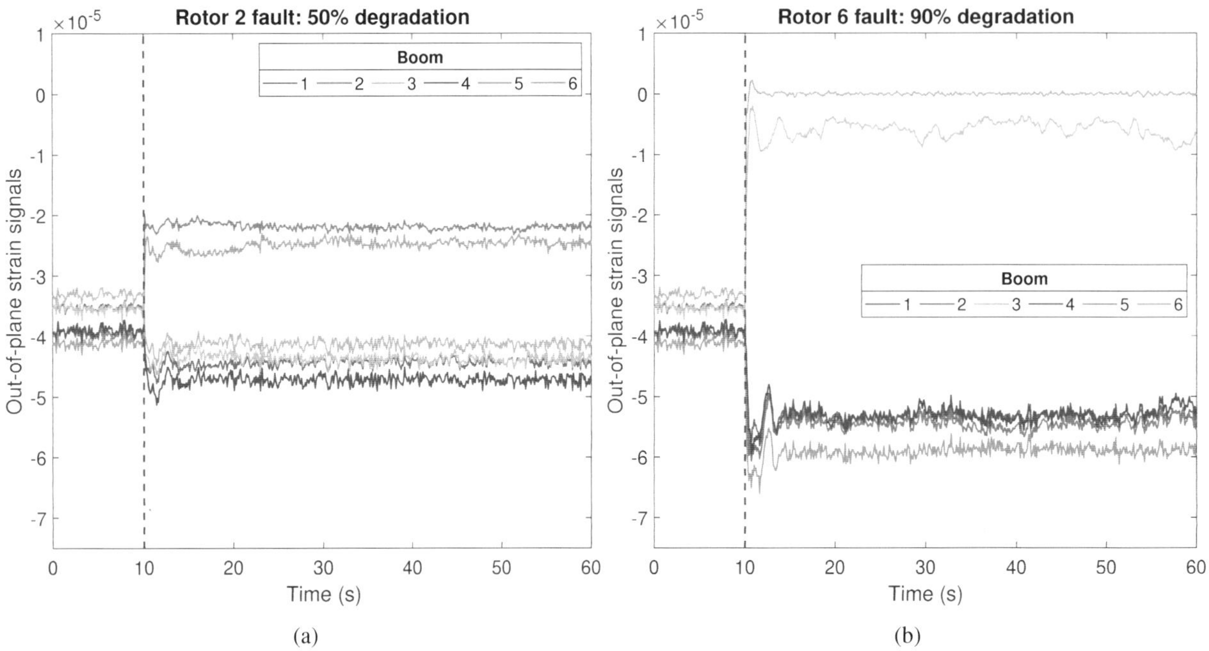

Figure 7: Time history of out-of- plane strain at each of the booms root under (a) 50 % degradation of rotor 2 (c) 90 % degradation of rotor 6

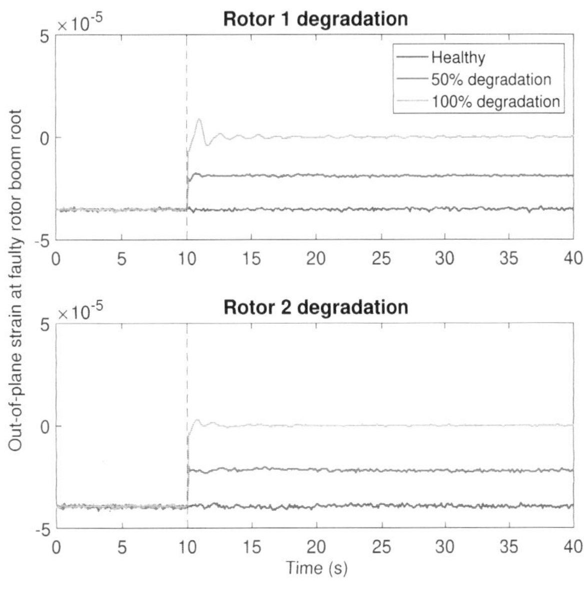

Figure 8: Out-of-plane strain signals at respective boom root with varying rotor fault magnitudes.

observed from the 6 local signals (out-of-plane strain at all boom roots) that the change in their absolute magnitude closely reflects the actual rotor speeds. Under different operating conditions, such as healthy flight with or without gusts, and different rotor fault scenarios, the statistical mean of signals will change relative to each other, forming some distinguishing patterns. These patterns can be classified via a pattern recognition algorithm, such as a single hidden layer neural Network to simultaneously detect and identify faults. If and when a fault is detected, the signal from the faulty rotor boom will be used for fault quantification by a simple linear regression framework. The workflow of the proposed rotor FDI technique is given in Fig. 9.

Neural Network Neural networks (NN), inspired from biological learning systems form a distributed information processing system consisting of neurons arranged in layers and executed in parallel. The neurons are nonlinear information processing elements and the interconnections between these neurons network are known as weights. These weights are learned through supervised training algorithms, where the training data contain the inputs and their corresponding output labels. NNs can compactly represent information, perform excellent classification, and can accommodate noise and uncertainty in data with carefully chosen features and regularization parameters. These characteristics make it attractive for the present application, where the aim is to develop a robust rotor FDI framework with signals affected by atmospheric disturbances encountered in actual flight.

From previous section, it has been observed that the values of out-of-plane strains at each boom root relative to each other can be a useful feature for a pattern recognition algorithm to determine the aircraft health status. To this end, a single-hidden layer is trained to output one of the 4 classes: healthy aircraft, rotor 1 fault, rotor 2 fault

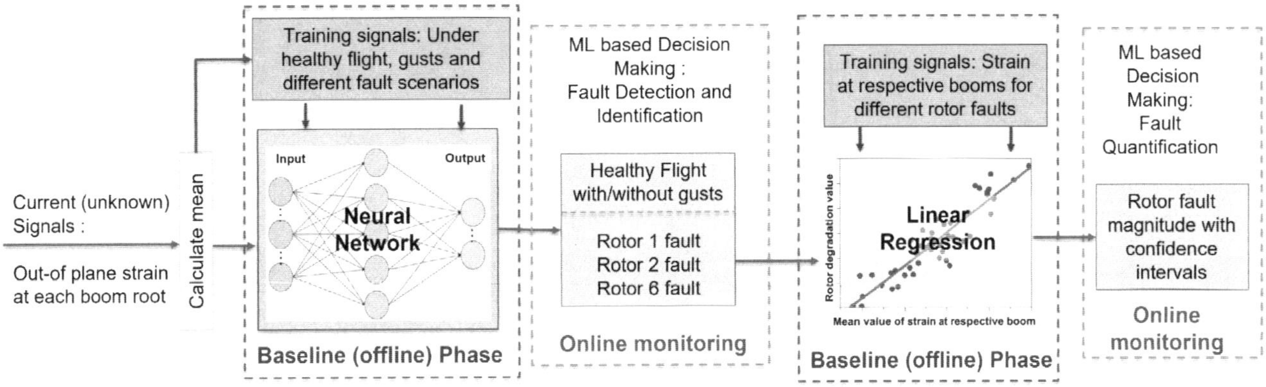

Figure 9: Machine learning based framework for probabilistic rotor fault detection, identification, and quantification

and rotor 6 fault with the mean value of the 6 strain signals over some time window as input. The input layer is denoted by $\mathbf{x}^T$ and the output layer is denoted by $\mathbf{h}(\mathbf{x})$ and is related by the following equation:

$$\mathbf{h}(\mathbf{x}) = \sigma\left(\mathbf{W_2}^T\left(\theta(\mathbf{W_1}^T\mathbf{x}) + \mathbf{B_1}\right) + \mathbf{B_2}\right) \quad (6)$$

where, $\theta(s)$ indicates the hyperbolic tangent activation function, and $\sigma(z)$ indicates the softmax activation function[1]. The weight matrices and bias vectors for the two layers denoted by $\mathbf{W_1}, \mathbf{W_2}$ and $\mathbf{B_1}, \mathbf{B_2}$ are determined in the baseline training phase by backpropagation learning techniques to minimize classification error. Using the cross-entropy error as cost function and softmax function in the output layer, the probabilities of each input element belonging to a output class can be obtained (Ref. 44).

The first class consists of healthy signals under severe turbulence with and without gusts (8.67 m/s with different directions). The rotor 1, 2, and 6 fault classes have been trained with the respective rotor fault signals under 10, 30, 50, 70, and 90% degradation (denoted as training data in Table 1). Note that the training data sets are divided into disjoint windows of 1 s. Each data-point consists of the mean of the 6 strain signals that are calculated over a single window and its corresponding true class label. The number of data-points should be balanced for the different classes to avoid classifier bias. The details of this NN are given in Table 2.

Linear Regression Model Assuming that the mean of the out-of-plane strain at the faulty rotor boom varies linearly with the rotor fault magnitude, a linear regression model is fitted (MATLAB function: *fitlm.m*) as follows:

$$y = \mathbf{x}\beta + \varepsilon \qquad \mathbf{x} = [1 \quad x] \qquad \beta = [\beta_1 \quad \beta_2]^T \quad (7)$$

where 'y' denotes the rotor degradation value, 'x' mean strain at faulty rotor boom and 'β' the vector of coefficients to be learnt from the training data.

The linear regression model is trained with the faulty training data denoted in Table 1. For rotor 1 faults, only the mean strain (calculated over window of 1 s) on boom 1 is used, with the labels being the true rotor degradation values. Similarly, for rotor 2 and 6 faults,the strain signals from boom 2 and 6 are used for training, respectively. The model outputs the predicted rotor fault magnitude, once the fault is identified by the NN along with the 95% confidence intervals according to Scheffé's method. The details of the trained linear regression model is given in Table 3.

Table 2: Details of the neural network training

Specifications	Value
Input Type	Mean of the out-of-plane strain at all boom roots
Input Layer Size	6
Training Function	Scaled Conjugate Gradient
Hidden Layer Size	3
Output Classes	4 (Healthy and 4 (Rotor 1,2 & 6 faults)
Cost Function	Cross-Entropy
Activation Function	1st layer - Hyperbolic Tangent Function , 2nd Layer - Softmax
Data Partition	Learning: Validation: Test = 50:25:25 % of Training Data
Performance	2.7×10^{-7}

RESULTS AND DISCUSSION

Current in-flight data streams have been used for online monitoring of aircraft health conditions. A window of 1

[1] Hyperbolic tangent activation function, $\theta(s) = \dfrac{e^s - e^{-s}}{e^s + e^{-s}}$

Softmax activation function, $\sigma(z_j) = \dfrac{e^{z_j}}{\sum_{i=1}^{C} e^{z_i}}$, C: no. of output classes,

$\mathbf{z} = \left(\mathbf{W_2}^T\mathbf{x}^{L-1} + \mathbf{B_2}\right)$

Table 3: Details of the linear regression training

Specifications	Value
Inputs	Mean of the out-of-plane strain at faulty rotor boom & rotor degradation value
Number of coefficients	2
R^2 fit	0.97
RMSE	4.9% rotor degradation
Output	Rotor degradation with confidence intervals

s of each of the 6 signals have been processed through workflow depicted in Fig. 10. To ensure a continuous decision making, the window is updated every 0.1 s. The computation time required to make a decision is less than 0.01 s, which is less than the window update time, making it suitable for real-time monitoring. A fault is detected when the neural network outputs any class other than healthy, and that output class determines which of rotor 1,2, or 6 has failed, enabling simultaneous fault detection and identification. Note that this decision is attached with a probability estimate which can help the operator to have a better insight of the fringe cases such as confusion between very mild degradation and strong gusts where the probability estimates of the two confusing classes will be about 50%, whereas in normal cases the confidence of a decision is generally more than 95%. After fault detection and identification, the linear regression model predicts the fault magnitude and its 95% confidence intervals using the strain from the boom on which the faulty rotor is mounted.

Indicative Results for Online Health Monitoring

Few indicative results for online monitoring of aircraft health with the current decision-making framework has been presented in Figs. 11a through 13b. The top subplots shown the mean of signals over a window of 1 s , updated every 0.1 s, that are used as input to the decision-making framework. The bottom subplots show the decision output by the framework, at each time window. The x-axis shows the time of flight in seconds in both the plots. Note that the online monitoring starts at 1 s to collect signals for the required time window. If the decision marker is shown in , the aircraft is healthy. Otherwise, the markers *, o, and x denote fault of rotor 1,2, and 6 respectively. The fault magnitude is shown by the y-axis in the bottom plots.

Figure 11a shows the correct health prediction for a healthy flight under light turbulence (a condition not used during training phase). Figures 11b and 12a, demonstrates that even under aggressive gusts of magnitude 8.67 m/s and 10 m/s, respectively which may cause significant change in the aircraft dynamics, the trained NN based on the static properties of the local signals correctly classifies it as a healthy aircraft.

Figures 12b to 13b, show indicative results for online monitoring under rotor faults. Note that the rotor fault magnitudes (20, 60 and 80% degradation) considered were not used in the training phase. In the top plots the time of fault commencement is shown in black vertical dashed lines. In the bottom plots the time of fault detection is shown in brown vertical dashed lines. A fault is always detected within 0.6 s of its commencement with longer time needed for mild degradation levels. In the bottom plots the true value of rotor degradation is shown by a grey horizontal line. It is observed that the true fault magnitude almost always lie within the 95% confidence intervals of the predicted values (shown by the vertical segments at each decision marker), denoting correct fault quantification. However, just after the fault detection and classification, depicted by the marker colour changing from green, the fault quantification becomes accurate only after the transient phase of the signals are over and they become steady (see Figs. 6b to 7b). The precision of fault quantification, i.e., the absolute difference between the predicted and true value is mostly less than 5% of rotor degradation in the steady state. The average range of 95% confidence intervals is 20 % rotor degradation.

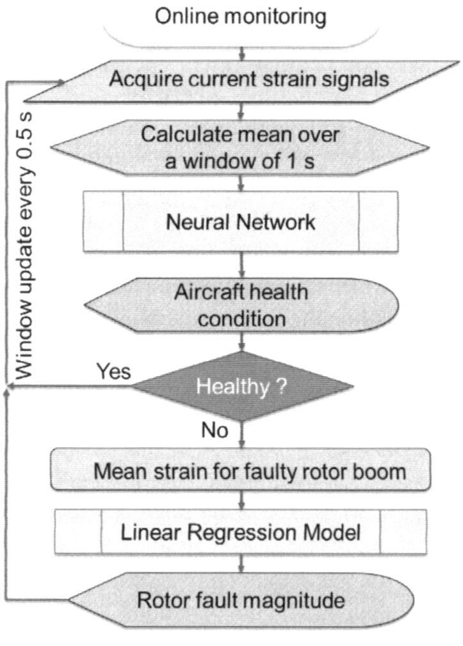

Figure 10: Flowchart for online monitoring of rotor faults

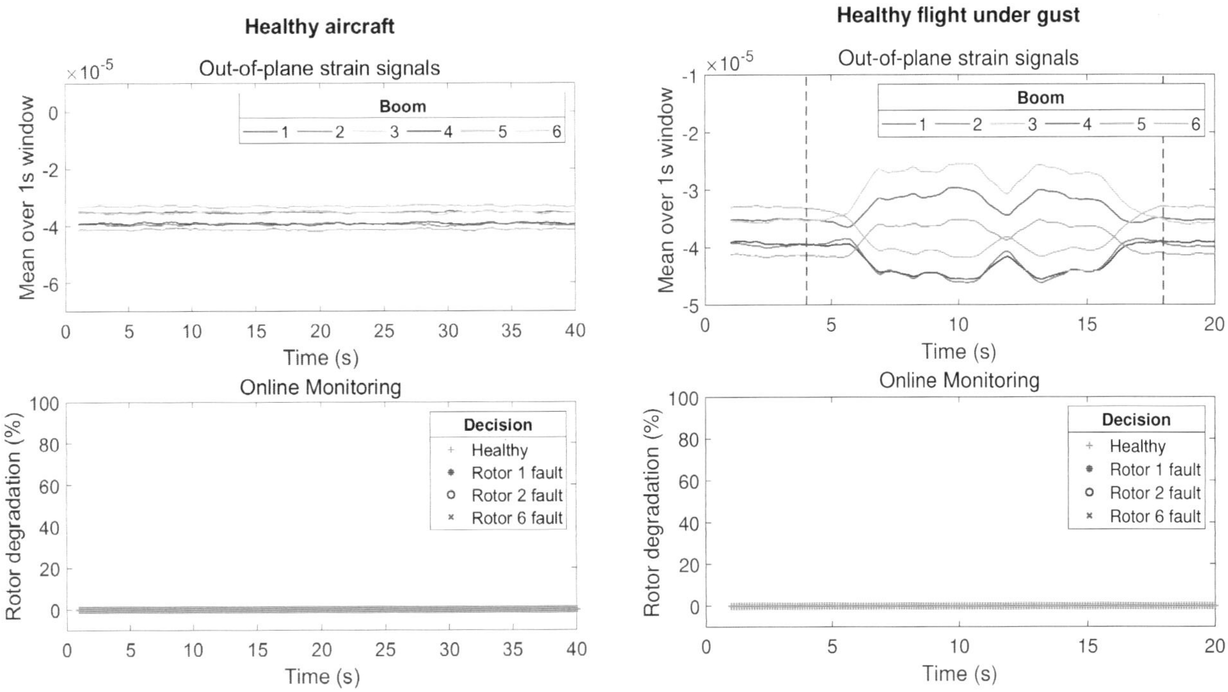

(a) (b)

Figure 11: Decision making in inspection phase under (a) healthy flight under light turbulence (b) healthy flight with gusts, $8.67(-0.57\hat{\imath}-0.57\hat{\jmath}+0.57\hat{k})$ m/s

Summary Results for the Performance

To ascertain the generalization capability of the neural network, data from various unknown operating conditions which have not been used to train it, have also been used in the inspection phase. These conditions include moderate and lights levels of turbulence in healthy flight, gusts of 5 m/s, 10 m/s and 3.46 m/s, and rotor faults of 20, 40, 60, 80, and 100% degradation (See Table 1). The performance of this framework on test data is summarized in Table 4. For each dataset, a number of decisions are made throughout the flight time with moving windows. The fault classification and quantification accuracy denotes the percentage of those decisions where the correct health condition is determined, and the if the true rotor degradation value lies in between the predicted 95% confidence intervals, respectively. The fault prediction error gives the average of the absolute error in fault magnitude estimation, and it is expressed in rotor degradation %.

It is observed that though the NN was trained with healthy data obtained under severe turbulence only, it gives perfect classification with healthy data under moderate and light turbulence levels too. For gusts with higher magnitude of wind velocity than used for training, the classification accuracy falls to 97.9%, otherwise it is very accurate. The rotor faults of 10 % or more degradation, have been classified accurately. The fault quantification accuracy is about 98%. This is due to the transient signals, where the fault magnitude is predicted lesser that the value obtained with steady signals. The absolute error i.e., the absolute difference between the predicted and true value of fault magnitude, averaged over all datasets and decision points is 4.2 % rotor degradation. The average time to detect a fault is less than 0.3 s. Mild rotor faults require more time to be detected, and faults over 50% degradation generally requires 0.1 s.

Comparison of the Unified Statistical Time-series Framework and the developed Machine-Learning framework

The machine-learning (ML) framework developed in this study has significant advantages over the previously established unified statistical time-series method. The most important is that the method developed in this study can distinguish aircraft response to gusts from that due to rotor faults, without any considerable extension of the nominal framework. The fault detection and identification in the ML framework is instantaneous and simultaneous at the commencement of fault. Whereas in the statistical time-series method, the fault identification only takes place after it has been compensated by the controller,

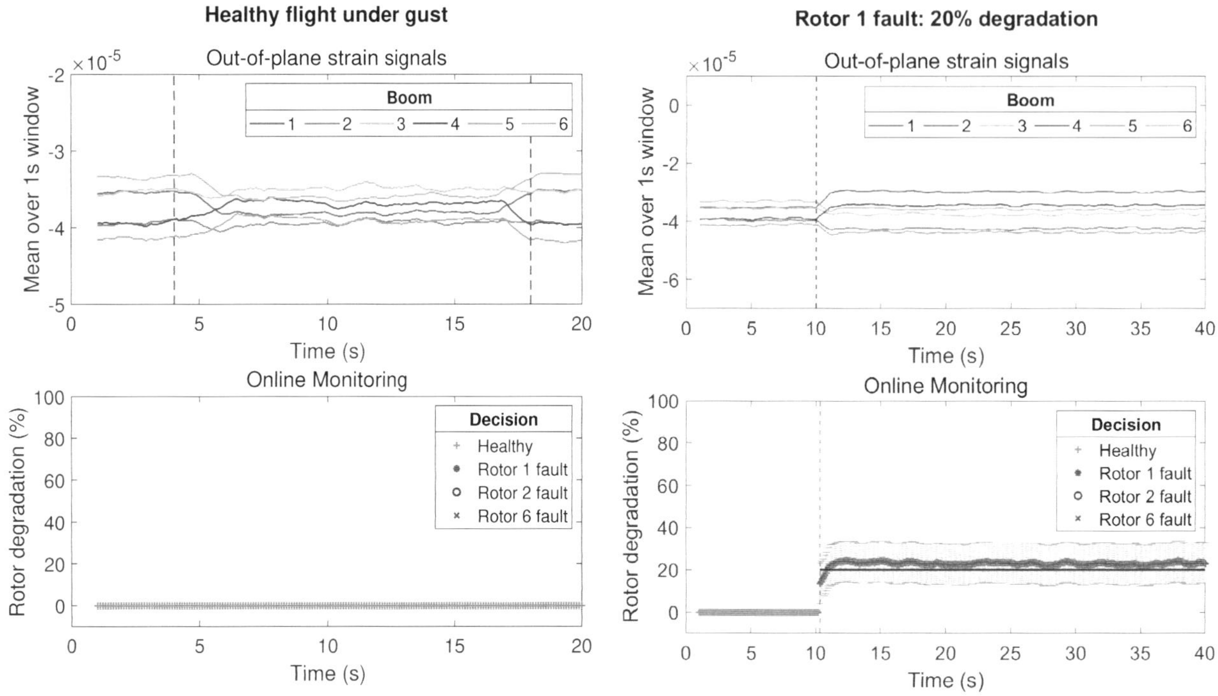

(a)

(b)

Figure 12: Decision making in inspection phase under (a) healthy flight with gusts, $10(-1\hat{i}+0\hat{j}+0\hat{k})$ m/s (b) 20 % degradation of rotor 1

Table 4: Fault Detection, Identification, and Quantification summary results

Aircraft state	Accuracy of Classification (%)	
Healthy	Severe Turbulence	100
	Moderate Turbulence	100
	Light Turbulence	100
Gusts	Wind velocity: 10 m/s	97.9
	Wind velocity: 8.67 m/s	100
	Wind velocity: 5 m/s	100
	Wind velocity: 3.46 m/s	100
Fault types	**Performance parameters**	
Rotor 1 faults	Classification accuracy (%)	100
	Fault Quantification accuracy (%)	99.6
	Fault prediction error (% degradation)	2.9
	Average time of detection (s)	0.22
Rotor 2 faults	Classification accuracy (%)	100
	Fault Quantification accuracy (%)	95.3
	Fault prediction error (% degradation)	6.2
	Average time of detection (s)	0.18
Rotor 6 faults	Classification accuracy (%)	100
	Fault Quantification accuracy (%)	99.8
	Fault prediction error (% degradation)	3.5
	Average time of detection (s)	0.22

because the identified models on which the decision-making depends on are valid with stationary signals only. Moreover, the signals need to be properly filtered and sampled at a higher sampling frequency ($F_s = 3333$ Hz). This ensures that the signals capture the aircraft and boom structural dynamics, for proper model identification to enable FDI by statistical time-series method. Longer signal window and more computation time is required for multiple binary hypothesis tests in statistical time-series method. The developed ML framework on the other hand, achieved high accuracy using signals' window of 1 s, with signals being sampled at only 10 Hz. However, in the ML framework, 6 signals are being collected and processed, for which special sensors needs to be mounted on each of the booms, whereas the statistical time-series method can work only with a single signal obtained from the aircraft IMU.

CONCLUSION

This paper introduces a machine-learning framework to detect, identify, and quantify rotor faults in a 2 kg gross weight hexacopter flying at 5 m/s under turbulence, and uncertainty while accurately rejecting disturbance due to gusts. In the baseline phase, commonly used ML algorithms such as a neural network and a linear regression

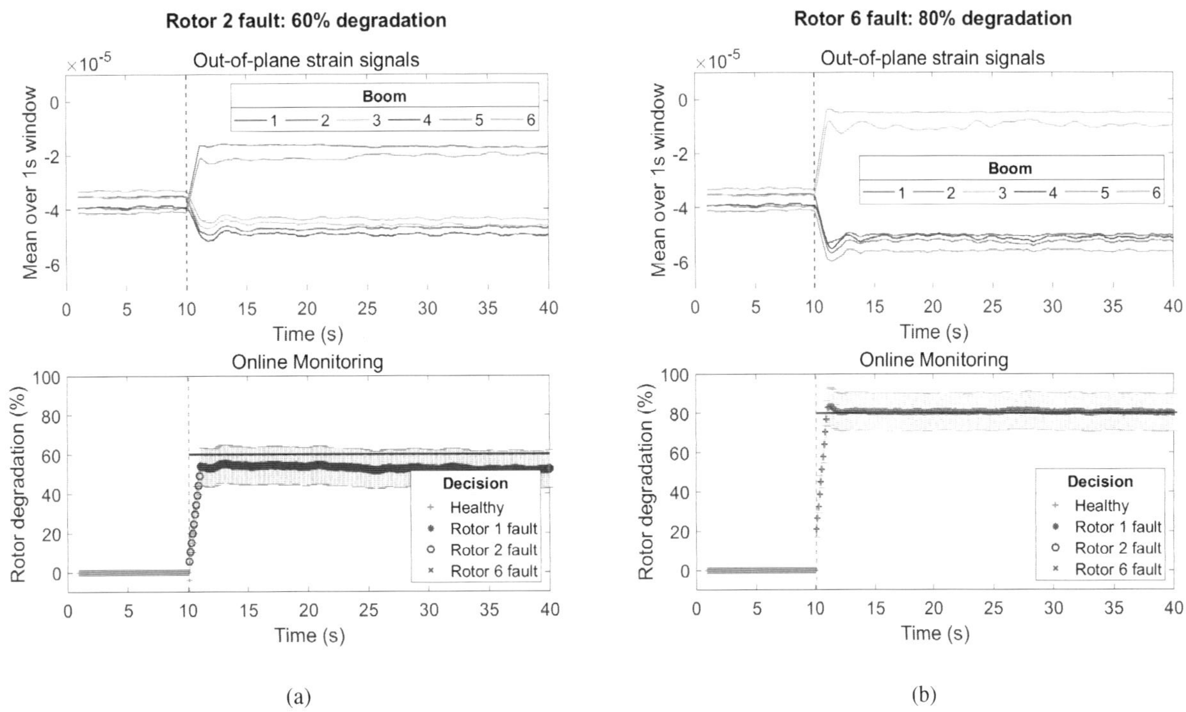

Figure 13: Decision making in inspection phase under (a) 60 % degradation of rotor 2 (b) 80 % degradation of rotor 6

model are trained on basis of static properties of the out-of-plane strain signals obtained from strain gauges placed at each of the booms' root. In its inspection phase, there are two distinct steps taking place within a probabilistic framework: step I involves simultaneous rotor fault detection and identification and step II involves fault quantification within the identified type of fault. Fault magnitude is considered to be continuous, involving an infinite number of potential fault magnitudes ranging from healthy to complete failure. The effectiveness of the method have been assessed via application to a number of "unmodeled" aircraft health conditions. The method is well-suited for online health monitoring of multicopters as demonstrated by the speed of decision-making, excellent accuracy with regard to determining gusts from different rotor faults, and precise fault magnitude estimation with uncertainty bounds.

The main conclusions drawn from this study are summarized as follows:

- The study – including the proof-of-concept application – has demonstrated that effective fault detection, fault mode identification, and fault magnitude estimation are possible based on judiciously selected sensors and signal analysis for extracting the embedded information pertaining to the aircraft health condition.
- Robust rotor fault detection, identification, and quantification for a hexacopter in forward flight have been achieved in real-time under different levels of turbulence, random gusts and uncertainty.
- Fault magnitude estimation at the nominal (point estimation) level is adequate, and at the probabilistic level it is accurate with close uncertainty bounds.
- The ML framework is better suited for real-time monitoring of aircraft health than the nominal unified statistical framework without any significant augmentation, provided proper instrumentation to monitor the strain on individual booms is available.
- A number of issues can be addressed in future work, e.g., the treatment of simultaneous multiple rotor fault scenarios, operation under varying flight speed and gross weight, and large uncertainties.

AUTHOR CONTACT

Airin Dutta	duttaa5@rpi.edu
Robert Niemiec	niemir2@rpi.edu
Fotis Kopsaftopolous	kopsaf@rpi.edu
Farhan Gandhi	fgandhi@rpi.edu

ACKNOWLEDGMENT

This work was supported by the U.S. Air Force Office of Scientific Research (AFOSR) grant "Formal Verification of Stochastic State Awareness for Dynamic Data-Driven Intelligent Aerospace Systems" (FA9550-19-1-0054) and Program Officer Dr. Erik Blasch.

REFERENCES

1. Booz, Allen, and Hamilton, "Urban Air Mobility (UAM) Market Study," Technical report, NASA, November 2018.

2. "Fast-Forwarding to a Future of On-Demand Urban Air Transportation," Technical report, October 2016.

3. "System Identification Modeling of a Small-Scale Unmanned Rotorcraft for Flight Control Design," Vol. 47, (1), January 2002, pp. 50–63. DOI: /10.4050/JAHS.47.50

4. Freddi, A., Longhi, S., Monteriù, A., and Prist, M., "Actuator Fault Detection and Isolation System for an Hexacopter," IEEE/ASME 10th International Conference on Mechatronic and Embedded Systems and Applications, August 2014. DOI: 10.1109/MESA.2014.6935563

5. D. Hyunchul Shim, H. J. K., and Sastry, S., "Control system design for rotorcraft-based unmanned aerial vehicles using time-domain system identification," IEEE International Conference on Control Applications, Anchorage, AK, 2000. DOI: 10.1109/CCA.2000.897539

6. Navarro, P., Cho, S., Rashid, A., Ruiz, A., and Bhandari, S., "Flight Testing, Data Collection, and System Identification of a Multicopter UAV," AIAA Modeling and Simulation Technologies Conference, Grapevine, TX, 2017. DOI: 10.2514/6.2017-1558

7. Serafini, J., Bernardini, G., Porcelli, R., and Masarati, P., "In-flight health monitoring of helicopter blades via differential analysis," *Aerospace Science and Technology*, Vol. 88, 2019, pp. 436–443. DOI: https://doi.org/10.1016/j.ast.2019.03.039

8. Heredia, G., and Ollera, A., "Sensor Fault Detection in Small Autonomous Helicopters using Observer/Kalman Filter Identification," IEEE International Conference on Mechatronics, Malaga, Spain, April 2009.

9. Waschburger, R., Paiva, H. M., e Silva, J. J. R., and Galvão, R. K. H., "Fault detection in a laboratory helicopter employing a wavelet-based analytical redundancy approach," Conference on Control and Fault-Tolerant Systems, Nice, France, 2010.

10. Schwartz, B. C., and Jones, D. L., "Quadratic and instantaneous frequency analysis of helicopter gearbox faults," *Mechanical Systems and Signal Processing*, Vol. 14, (4), 2000, pp. 579–595.

11. Williams, W. J., and Zalubas, E. J., "Helicopter transmission fault detection via time-frequency, scale and spectral methods," *Mechanical Systems and Signal Processing*, Vol. 14, (4), 2000, pp. 545–559.

12. Gopalratnam, G., Zorn, C., and Koch, A., "Multisensor Data Fusion for Sensor Failure Detection and Health Monitoring," AIAA Guidance, Navigation, and Control Conference and Exhibit, San Francisco, CA, August 2005.

13. Cox, J. R., Arnold, S., and Anusonti-Inthra, P., "Rotor Blade Fault Detection through Statistical Analysis of Stationary Component Vibration," 52nd AIAA/SAE/ASEE Joint Propulsion Conference, July 2016. DOI: 10.2514/6.2016-4642

14. Lu, F., Li, Z., Huang, J., and Jia, M., "Hybrid State Estimation for Aircraft Engine Anomaly Detection and Fault Accommodation," *AIAA Journal*, Vol. 58, (4), 2020, pp. 1748–1762. DOI: 10.2514/1.J059044

15. Morel, H., Ouladsine, M., Krysinski, T., and Brun-Picard, D., "Defect detection and tracing on helicopter rotors by artificial neural networks," IEEE Advanced Process Control Applications for Industry Workshop, Vancouver, Canada, 2005.

16. Ganguli, R., Chopra, I., and Haas, D. J., "Helicopter Rotor System Fault Detection Using Physics-Based Model and Neural Networks," *AIAA Journa*, Vol. 36, (6), May 1998, pp. 1078–1086.

17. Gino Iannace, G. C., and Trematerra, A., "Fault Diagnosis for UAV Blades Using Artificial Neural Network," *Robotics*, Vol. 8, (3), 2019. DOI: 10.3390/robotics8030059

18. Ganguli, R., "Health Monitoring of a Helicopter Rotor in Forward Flight Using Fuzzy Logic," *AIAA Journal*, Vol. 40, (12), December 2002.

19. Pawar, P. M., and Ganguli, R., "Fuzzy-Logic-Based Health Monitoring and Residual-Life Prediction for Composite Helicopter Rotor," *Jour-*

nal of Aircraft, Vol. 44, (3), 2007, pp. 981–995. DOI: 10.2514/1.26495

20. Chen, Z., and Yang, Y., "Fault diagnostics of helicopter gearboxes based on multi-sensor mixtured hidden markov models," *ASME Journal of Vibration and Acoustics*, Vol. 134, (3), 2012.

21. Wang Min, H. N.-q., and Guo-jun, Q., "A method for rule extraction based on granular computing: Application in the fault diagnosis of a helicopter transmission system," *Journal of Intelligent and Robotic Systems*, Vol. 71, 2013, pp. 445–455.

22. Ochoa, C. A., and Atkins, E. M., "Multicopter Failure Diagnosis through Supervised Learning and Statistical Trajectory Prediction," 2018 AIAA Information Systems-AIAA Infotech @ Aerospace, Kissimmee, Florida, January 2018. DOI: 10.2514/6.2018-1636

23. Baskaya, E., Bronz, M., and Delahaye, D., "Fault detection diagnosis for small UAVs via machine learning," 2017 IEEE/AIAA 36th Digital Avionics Systems Conference (DASC), 2017.

24. Bondyra, A., Gasior, P., Gardecki, S., and Kasiński, A., "Fault diagnosis and condition monitoring of UAV rotor using signal processing," , 09 2017. DOI: 10.23919/SPA.2017.8166870

25. Qi, X., Theillol, D., Qi, J., Zhang, Y., and Han, J., "A Literature Review on Fault Diagnosis Methods for Manned and Unmanned Helicopters," International Conference on Unmanned Aircraft Systems, May 2013.

26. Fassois, S., and Kopsaftopoulos, F., *New Trends in Structural Health Monitoring*, Springer, January 2013, Chapter Statistical Time Series Methods for Vibration Based Structural Health Monitoring, pp. 209–264. DOI: 10.1007/978-3-7091-1390-5

27. Kopsaftopoulos, F. P., and Fassois, S. D., "Scalar and Vector Time Series Methods for Vibration Based Damage Diagnosis in a Scale Aircraft Skeleton Structure ," *Journal of Theoretical and Applied Mechanics*, Vol. 49, (4), 2011.

28. Samara, P. A., Fouskitakis, G. N., Sakellariou, J. S., and Fassois, S. D., "A Statistical Method for the Detection of Sensor Abrupt Faults in Aircraft Control Systems," *IEEE Transactions on Control Systems Technology*, Vol. 16, (4), July 2008, pp. 789–798. DOI: 10.1109/TCST.2007.903109

29. Kopsaftopoulos, F. P., and Fassois, S. D., "A vibration model residual-based sequential probability ratio test framework for structural health monitoring," *Structural Health Monitoring*, Vol. 14, (4), 2015, pp. 359–381.

30. Dimogianopoulos, D. G., Hios, J. D., and Fassois, S. D., "FDI for Aircraft Systems Using Stochastic Pooled-NARMAX Representations: Design and Assessment," *IEEE Transactions on Control Systems Technology*, Vol. 17, (6), November 2009, pp. 1385–1397.

31. Dimogianopoulos, D., Hios, J., and Fassois, S., "Aircraft engine health management via stochastic modelling of flight data interrelations," *Aerospace Science and Technology*, Vol. 16, (1), 2012, pp. 70–81. DOI: 10.1016/j.ast.2011.03.002

32. Zhao, Z., Quan, Q., and Cai, K.-Y., "A Health Evaluation Method of Multicopters Modeled by Stochastic Hybrid System," *Aerospace Science and Technology*, Vol. 68, September 2017, pp. 149–162. DOI: 10.1016/j.ast.2017.05.011

33. Dutta, A., McKay, M., Kopsaftopoulos, F., and Gandhi, F., "Statistical residual-based time series methods for multicopter fault detection and identification," *Aerospace Science and Technology*, Vol. 112, 2021, pp. 106649. DOI: https://doi.org/10.1016/j.ast.2021.106649

34. Dutta, A., McKay, M., Kopsaftopoulos, F., and Gandhi, F., "Rotor Fault Detection and Identification on a Hexacopter under Varying Flight States Based on Global Stochastic Models," Vertical Flight Society 76th Annual Forum, Online (due to COVID-19), October 2020.

35. Dutta, A., McKay, M., Kopsaftopoulos, F., and Gandhi, F., "Fault Detection and Identification for Multirotor Aircraft by Data-Driven and Statistical Learning Methods," Electric Aircraft Technologies Symposium (EATS), Indianapolis, IN, August 2019.

36. Dutta, A., McKay, M., Kopsaftopoulos, F., and Gandhi, F., "Rotor Fault Detection and Identification for a Hexacopter Based on Control and State Signals via Statistical Learning Methods," Vertical Flight Society 76th Annual Forum, Online (due to COVID-19), October 2020.

37. Dutta, A., Niemiec, R., Kopsaftopoulos, F., and Gandhi, F., "Unified Statistical Framework for Rotor Fault Diagnosis on a Hexacopter via Functionally Pooled Stochastic Models," Vertical Flight So-

ciety 77th Annual Forum, Online (due to COVID-19), May 2021.

38. Peters, D., and He, C., "A Finite-State Induced Flow Model for Rotors in Hover and Forward Flight," American Helicopter Society 43rd Annual Forum, St. Louis, MO, May 1987.

39. McKay, M., Niemiec, R., and Gandhi, F., "Post-Rotor-Failure-Performance of a Feedback Controller for a Hexacopter," American Helicopter Society 74th Annual Forum, Phoenix, AZ, May 2018.

40. Hakim, T. M. I., and Arifianto, O., "Implementation of Dryden Continuous Turbulence Model into Simulink for LSA-02 Flight Test Simulation," Journal of Physics: Conference Series 1005(2018) 012017, August 2018.

41. Davoudi, B., Taheri, E., Duraisamy, K., Jayaraman, B., and Kolmanovsky, I., "Quad-Rotor Flight Simulation in Realistic Atmospheric Conditions," *AIAA Journal*, Vol. 58, (5), 2020, pp. 1992–2004. DOI: 10.2514/1.J058327

42. Palomaki, R. T., Rose, N. T., van den Bossche, M., Sherman, T. J., and Wekker, S. F. J. D., "Wind Estimation in the Lower Atmosphere Using Multirotor Aircraft," *Journal of Atmospheric and Oceanic Technology*, May 2017. DOI: 10.1175/JTECH-D-16-0177.1

43. Wang, B. H., Wang, D. B., Ali, Z. A., Ting, B. T., and Wang, H., "An Overview of Various Kinds of Wind Effects on Unmanned Aerial Vehicle," *Journal of Measurement and Control*, Vol. 52, (7-8), September 2019, pp. 731–739. DOI: 10.1177/0020294019847688

44. Lan, H., "The Softmax Function, Neural Net Outputs as Probabilities, and Ensemble Classifiers," .

APPENDIX

A. Signals

The data used in this study are obtained from simulation rather than experiments and therefore the sensor signals need to be calculated from the available aircraft states.

Generally, Inertial Measurement Units (IMUs) are composed of a 3-axis accelerometer and a 3-axis gyroscope and outputs the body accelerations (x, y, and z acceleration) and the angular rates (roll, pitch, pitch and yaw rates), which can be determined from 12 rigid body states are defined in Eq. 1 as follows:

$$\begin{aligned} \text{Body accelerations} &= \begin{bmatrix} \dot{u} & \dot{v} & \dot{w} \end{bmatrix}^T \\ \text{Angular rates, } \omega &= \begin{bmatrix} p & q & r \end{bmatrix}^T \end{aligned} \tag{8}$$

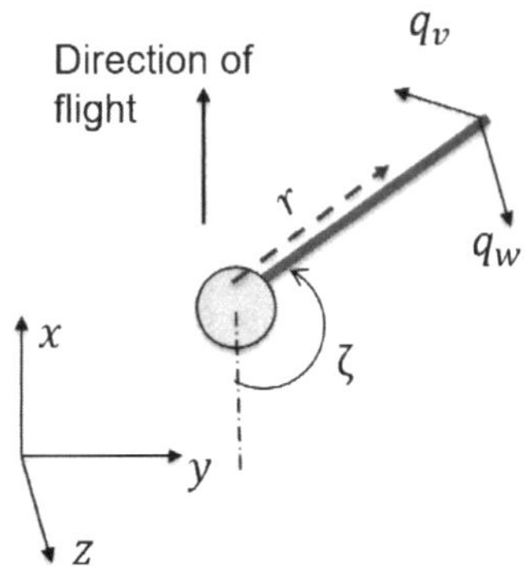

Figure 14: Position of a boom with respect to the hub and its deformations

These signals are referred to as remote signals.

The sensors mounted on the different locations on the individual booms, such as strain gauges and accelerometers are referred to as the local signals. These can be computed from the individual booms modal deformation states and shape functions (Eqs. 5 and 3) using the following expressions:

$$\begin{aligned} \text{Total deformation, } q &= \begin{cases} \sum_{i=1}^{2} \phi_{w_i}(l)\eta_{w_i} & \text{Out-of-plane} \\ \sum_{i=1}^{2} \phi_{v_i}(l)\eta_{v_i} & \text{In-plane} \end{cases} \\ \text{Strain, } \varepsilon &= \begin{cases} \sum_{i=1}^{2} \phi''_{w_i}(l)\eta_{w_i} \times a/2 & \text{Out-of-plane} \\ \sum_{i=1}^{2} \phi''_{v_i}(l)\eta_{v_i} \times b/2 & \text{In-plane} \end{cases} \end{aligned} \tag{9}$$

where, a and b are the width and height of the cross-section of the boom, respectively. l is the distance from the boom root where the sensor has been placed, normalized by the boom length.

Table 5: Boom properties

Parameters	Value
Boom Length (L)	0.2617 m
Material	Aluminium
Cross-section	Hollow square
Outer dimension	0.0156 m
Inner dimension	0.0130 m
Flexural Rigidity	179 Nm^2

The accelerometer readings on each of the boom can be calculated as:

$$\text{Acceleration} = \ddot{r} + \omega \times \dot{r} + \dot{\omega} \times r + \omega \times \omega \times r \tag{10}$$

where,

Distance of accelerometer from the hub, r

$$= \begin{bmatrix} -L\cos\zeta \\ L\sin\zeta \\ -d \end{bmatrix} + \begin{bmatrix} -\Sigma_{i=1}^{2}\phi_{v_i}(l)\eta_{v_i}\sin\zeta \\ -\Sigma_{i=1}^{2}\phi_{v_i}(l)\eta_{v_i}\cos\zeta \\ \Sigma_{i=1}^{2}\phi_{w_i}(l)\ddot{\eta}_{w_i} \end{bmatrix}$$

Here, ζ is the azimuth angle of the boom (See Fig. 14), L is the length of the boom and d is the vertical position of the boom from the center-of-gravity of the aircraft. The boom properties are given in Table 5.

Note that these derivations are shown for a single time instant, t. Repeating this computations for the entire range of time will generate the time-series sensor data.

B. Gusts

The gusts considered in this study, along with their magnitude and direction are tabulated below.

Table 6: Gusts Label

Label	Magnitude (m/s)	Direction
1	8.67	$0.57\hat{i}+0.57\hat{j}+0.57\hat{k}$
2	8.67	$0.57\hat{i}+0.57\hat{j}-0.57\hat{k}$
3	8.67	$0.57\hat{i}-0.57\hat{j}+0.57\hat{k}$
4	8.67	$-0.57\hat{i}+0.57\hat{j}+0.57\hat{k}$
5	8.67	$-0.57\hat{i}-0.57\hat{j}+0.57\hat{k}$
6	8.67	$0.57\hat{i}-0.57\hat{j}-0.57\hat{k}$
7	8.67	$-0.57\hat{i}+0.57\hat{j}-0.57\hat{k}$
8	8.67	$-0.57\hat{i}-0.57\hat{j}-0.57\hat{k}$
9	3.46	$0.57\hat{i}+0.57\hat{j}+0.57\hat{k}$
10	3.46	$0.57\hat{i}+0.57\hat{j}-0.57\hat{k}$
11	3.46	$0.57\hat{i}-0.57\hat{j}+0.57\hat{k}$
12	3.46	$-0.57\hat{i}+0.57\hat{j}+0.57\hat{k}$
13	3.46	$-0.57\hat{i}-0.57\hat{j}+0.57\hat{k}$
14	3.46	$0.57\hat{i}-0.57\hat{j}-0.57\hat{k}$
15	3.46	$-0.57\hat{i}+0.57\hat{j}-0.57\hat{k}$
16	3.46	$-0.57\hat{i}-0.57\hat{j}-0.57\hat{k}$
17	5	$1\hat{i}+0\hat{j}+0\hat{k}$
18	5	$-1\hat{i}+0\hat{j}+\hat{k}$
19	5	$0\hat{i}+1\hat{j}+0\hat{k}$
20	5	$0\hat{i}-1\hat{j}+0\hat{k}$
21	5	$0\hat{i}+0\hat{j}+1\hat{k}$
22	5	$0\hat{i}+0\hat{j}-1\hat{k}$
23	10	$1\hat{i}+0\hat{j}+0\hat{k}$
24	10	$-1\hat{i}+0\hat{j}+\hat{k}$
25	10	$0\hat{i}+1\hat{j}+0\hat{k}$
26	10	$0\hat{i}-1\hat{j}+0\hat{k}$
27	7	$0\hat{i}+0\hat{j}+1\hat{k}$
28	7	$0\hat{i}+0\hat{j}-1\hat{k}$

Candidate VMS Architecture Approaches to Support Future AH-64 Apache Attack Helicopter Requirements

Bryan C. H. Chu
Sr. GN&C Engineer
The Boeing Company
Mesa, AZ, Country

Gary Klein
Associate Technical Fellow
The Boeing Company
Mesa, AZ, USA

Russell Enns
Technical Fellow
The Boeing Company
Mesa, AZ, USA

Tom Gollem
Sr. Avionics Systems Design Engineer
The Boeing Company
Mesa, AZ, Country

ABSTRACT

Multiple Vehicle Management System (VMS) architectures are presented for the AH-64 attack helicopter addressing several demanding and possibly conflicting future requirements, including level 1 handling qualities, operating in degraded visual environments (DVE), autonomy, high-speed operation, and multiple vehicle coordinated operations. Architecture tradeoffs are performed with the understanding that the current AH-64 flight control system is mechanical with an electrical partial authority flight control system, but also possesses a non-redundant full authority fly-by-wire emergency backup system. The various architectures are informally qualitatively assessed as to how they satisfy the requirements. They are also assessed with respect to their relative costs, both for typical costs such as design, recurring, operating and sustainment, and training as well as other indirect costs such as size and weight. A methodology is developed to compare and contrast the architectures using informal qualitative-based scores based on the authors' experiences. The results suggest that there are several possibly cost effective options, possibly with a higher potential return on investment (ROI) than a conventional fly-by-wire (FBW) architecture. Some of these solutions are also incremental in nature, thereby providing more flexibility than the conventional FBW system.

INTRODUCTION

The AH-64's capabilities have continued to advance over the last 30 years but, while more and more enhancements have been added from a mission capability perspective, very little of the flight control system has changed or evolved. As a result, crew workload continues to increase without any alleviation from the features that modern flight control systems afford. While this has been acceptable to date, the continued evolution of requirements points more and more to the need for a next generation advanced flight control system. As one example, a requirement to not only operate in degraded visual environments (DVE) but to also exploit the tactical advantage of DVE points to the necessity for a much more stable and less decoupled control system. As another example, a requirement for autonomy points to the necessity of a full authority digital flight control system that has some level of redundancy. Other features such as multiple vehicle coordinated operations, and perhaps even higher speed compound designs, also point to the need for an improved flight control system.

While fly-by-wire is one possibility to meet these requirements, there is a long history that has made that option more daunting than one may want to pursue if other options exist. The AH-64's underlying flight control system is unique in some respects, particularly with it possessing a full authority single strand fly-by-wire (FBW) emergency back up control system to augment its hydraulically-boosted mechanical flight control system with stability augmentation. This uniqueness provides a path to satisfy requirements using flight control system architectures other than the traditional triplex FBW often seen. In this paper we examine such architectures and compare and contrast them, not only with respect to the requirements they satisfy, but also with respect to their relative costs. Cost includes not only direct cost such as design, verification and validation (V&V), operating and sustainment (O&S), recurring, and training but also indirect costs such as size, weight, and risk.

The Need for a Next Generation AH-64 Flt Ctrl System

The need for a next generation flight control system is driven by several new and evolving requirements. These requirements are quite varied and include improved handling qualities/reduced workload, operating in DVE, optimizing vehicular performance, improved mission task effectiveness, reduced operating and sustainment costs, improved safety,

Presented at the VFS Modern Ops. with Future Integrated VMS Technical Meeting, West Chester, PA, Aug. 19-20, 2021.

providing the ability to add more autonomous capabilities, multi-vehicle coordinated operations, and higher speed operation.

In addition to the new and evolving requirements, there are other well-known existing requirements that need to be met such as safety of flight reliability and mission reliability. The goal of this study is to examine these requirements in the context of the various next generation flight control system candidates to see which are best met for these varied and possibly conflicting requirements.

Helicopter FBW History

One well established path to providing the aforementioned capabilities for the Apache is to convert the existing mechanical flight control system to fly-by-wire. However, the history of FBW helicopter flight control systems has been a long slow journey. Unlike many new technologies that fairly rapidly transition from introduction to pervasive adoption, this has not been the case for FBW helicopters. The issue is that not only is the recurring cost as a percentage of vehicle cost high (compared to much larger fixed wing aircraft), the quantity of vehicles sold is often lower, and the ability to eventually recoup the investment is either not well appreciated or requires long term commitment.

The first fly by wire helicopter was developed in the 1970s with the BV-347 demonstrator [Ref. 1]. It was developed purely for technology demonstration without any plans for production. The 1980s witnessed new digital flight control R&D with the Advanced Digital/Optical Control System (ADOCS) program on a UH-60 [Ref. 2] as well as attempts to put the technology into production. Specifically the MDHC-Bell LHX program demonstrated a FBW system on AV-05, a non-production AH-64 aircraft. However, it lost out to the Boeing-Sikorsky LHX competitor, which advanced helicopter FBW evolution one step further with the RAH-66 Comanche. Unfortunately, due to ever evolving requirements and consequent cost growth, the Comanche program was ultimately cancelled. More FBW demonstrators ensued through the 1990s including the USG's RASCAL program [Ref. 3] and the MDHC FBWDS/RPA programs [Ref. 4].

It was not until the late 1990s that the first production FBW helicopter, the NH-90, came to fruition, finally entering production in the 2000's [Ref. 5]. Perhaps the reason for its success was that it took the form of a nontraditional FBW architecture. While consisting of four redundant channels, two of the channels were digital computers while the other two were analog computers. By introducing analog computers into the redundancy scheme, the amount of formal qualification testing was significantly less than a traditional digital triplex system.

Even after the NH-90 went into production, FBW did not rapidly make its way into most helicopters. Sikorsky and Boeing started a collaborative production FBW program for the UH-60 Blackhawk and AH-64 Apache around 2004. The customer halted the Boeing AH-64 effort within a year once the overall costs were better determined and other technologies such as mission equipment upgrades out-prioritized the flight control system based on the cost-benefit assessment at that time. The UH-60 effort continued for several years but was ultimately cancelled while in flight test due to a variety of reputed reasons including cost growth, processor obsolescence, and complexity. Meanwhile, Sikorsky was also working on a similar FBW system for the S-92 (Canadian Coast Guard) and the CH-53K shortly thereafter. The S-92 FBW effort was more successful than the UH-60, presumably from lessons learned by Sikorsky, and became the second production FBW system to enter service. The next FBW effort of note was the Bell Relentless 525 [Ref. 6], a medium-lift helicopter that was first announced in 2012 and is just now approaching final certification.

The time that all of the successful FBW efforts have taken to get into production points to high development, verification, validation, and certification costs as somewhat presciently noted by Ray Prouty [Ref. 7]. In fact, history has shown that, while highly touted for the capabilities it offers [Ref. 1], the introduction of FBW into helicopters has been very slow, even slower than for fixed wing which, in itself, was slow. It has been slow because of both development and certification costs as well as the recurring (equipment cost per aircraft) costs. A large driver of this cost is the need for triplex (or more) redundancy; the recurring cost growing linearly with redundancy but the certification cost increasing more exponentially. That cost has grown even more because of the increased certification scrutiny from aviation governing authorities ensuring compliance with industry processes such as DO-178. While laudable from a process and compliance perspective (though arguably overly risk-averse), the cost to meeting compliance and certification liability is now rapidly outpacing other elements of the development costs.

FBW becomes more viable as the vehicle size increases where the ultimate cost for the vehicle is higher, and hence percentage of cost for FBW is lower. Moreover, the benefits are greater on larger vehicles as well (which have often ended up being more about weight savings than handling qualities (HQ) and capability improvements). Unfortunately, the AH-64 is not a relatively large vehicle and, consequently, weight savings has not been as compelling a case.

For these reasons, either new approaches to reducing traditional FBW costs are needed, or alternatives to a traditional triplex FBW system are sought. Most recently, the H-47 helicopter, facing similar challenges, has opted to pursue an approach called the Active Parallel Actuation System (APAS) that provides many of the benefits of FBW but at less cost [Ref. 8, 9]. This paper develops and analyzes several alternatives for the AH-64, including both FBW and APAS but also some other novel approaches, in light of the previously described emerging requirements for the vehicle.

Past AH-64 History

The AH-64 Apache helicopter has served as the attack helicopter for the US Army, and the pre-eminent attack helicopter in the world, spanning a period of well over 30 years. Over this time the aircraft has evolved dramatically, most notably in the avionics mission equipment package, witness the 28 to 1 force multiplier in lethality from the A to the D model [Ref. 10] and [Ref. 11], but also in pretty much every other aspect except the flight control system. The AH-64's flight control system is largely the same as it was at its onset, with only small additions of hold modes, a Translational Rate Command (TRC) mode, and improved turn coordination from a functional point of view, and upgrades to the inertial sensing and flight computers from a hardware perspective.

Much of the limitation in functionality improvements has been due to the Apache's underlying flight control system being mechanical with a limited authority stability control augmentation system (SCAS). That limits the flight control system's capability to rate damping for improved stabilization, some turn coordination, hold modes (altitude, position, velocity, attitude) up until the point where the SCAS saturates, and a TRC mode (velocity command/position hold) at very low velocities.

Awareness of such limitations has prompted multiple attempts to improve the AH-64's flight control system. Various FBW demonstration efforts have shown the art of the possible, starting with AV-05 as part of the LHX program in the late 1980s. This was followed by the FBWDS and RPA programs in the 1990s, and the VITAL and HACT programs in the early 2000s. With the realization of the value of these technologies, an effort to develop a production FBW system for the AH-64 began in 2004 but the customer halted the program when the full cost to productionize the fleet was better understood. This then led to a program to add Modern Control Laws (MCLAWS) to the existing flight control system (albeit with SCAS authority increased from +/-10% to +/-20%). However, that program was also stopped when the full development, flight test and refurbishment costs were realized.

In summary, a FBW system provides the AH-64 with all the desired benefits but at a cost beyond what has been deemed to be affordable, whereas an increased SCAS solution provided much less benefits but still at a hefty cost/benefit ratio. This realization motivated the need to find the sweet spot, providing a system with much or all of the benefit but at less cost than FBW. Unlike all other non-FBW helicopters, the AH-64 is uniquely positioned to lend itself to alternative approaches. Specifically, the existing Apache has hydraulically-boosted single path mechanical flight controls with SCAS and a single-strand FBW Backup-Control System (BUCS) that becomes active when the mechanical controls are either jammed or severed. This novel architecture makes possible a variety of FBW system designs, one previously published and the others published here for the first time.

The one previous published concept, developed in the 2007-2008 time frame, was to keep the existing control but with a (de)coupler used between the pilot stations and the main servo actuators [Ref. 12]. The system would nominally operate in BUCS mode with the coupler in the decoupled state. A failure of the single strand system (albeit redundant sensors) would then result in re-coupling the two stations to the mechanical control linkage so that the pilot could then fly the system through the mechanical flight control system. This provided a single strand FBW system, with all the benefits of FBW, but without the cost of triplex redundancy. It required excellent robust fault detection, but also needed to account for servo actuator command transients (along with vehicular motion transients) that would ensue as the system recoupled.

Before fully developing this idea, it was discounted because of unproven concerns about aircraft transients during recoupling after a failure and a lack of priority to fund analyzing the merits of the concerns and whether they could be addressed. However, over time the value of next generation flight control systems to aid with a variety of important military objectives has become more and more recognized. Thus, it is hopeful that a next generation flight control systems architecture for the AH-64 will be more favorably considered.

AH-64 Flight Control System Possibilities

Based on the previous discussions and taking advantage of the AH-64's single-strand FBW Backup-Control System (BUCS), many different architectures are possible. This paper explores several of them including:

- Traditional Fly-By-Wire
- Active Parallel Actuation System (APAS)
- APAS with Coupler
- Dissimilar Redundancy Extensible FBW

Paper Organization

This paper delves into the details of these various architectures, the various requirements that need to be satisfied, how well the various architectures do in satisfying those requirements, and the costs associated with each of the architectures. Specifically, the first section looks at the various anticipated requirements. The next section of this paper explores several different candidate architectures. With the requirements described and the architectures described, the paper then informally and qualitatively discuss how each of the various architectures addresses the various requirements. An informal and qualitative discussion of the costs associated with each architecture follows. The paper then takes an approach to aggregating these assessments. These informal results are then summarized for conciseness, including a plot showing the relative value versus cost of each architecture, and conclusions drawn.

REQUIREMENTS

The need for a next generation AH-64 flight control systems is driven by several emerging and anticipated future requirements. The following discusses each of these requirements.

- Improved handling qualities/reduced workload – The need for both improved handling qualities and reduced workload are a recurring theme for the AH-64. In summary, it is desired that the Apache have level 1 HQ but the aircraft is closer to level 2 HQ and flight control system improvements are the only way to improve on that. More important than just a handling qualities level, there is the realization that the pilot and co-pilot's tasks have continually increased over the last 30 years, to the point where they are becoming task saturated. By providing a next generation flight control system that can support advanced task tailored control laws with active inceptor controls that provide tactile cueing, the workload can be dramatically reduced. For example, the AVMS program saw a 34% decrease in Task Load Index by applying advanced control laws with tactile cueing [Ref. 13].
- For attack helicopters, improved handling qualities and reduced workload also have to include factors that traditional ADS-33 [Ref. 14] Handling Qualities evaluations do not focus on. Specifically, a basic attack helicopter operational requirement is that pilots need to maneuver the aircraft as rapidly as possible. These maneuvers include either engaging (i.e., acquiring, targeting and firing the weapon) the enemy or evading the enemy (i.e. preventing or breaking the enemy's weapons sensor lock or to maneuver the aircraft such that the trajectory exceeds the enemy's weapon kinetic capability to hit the aircraft). ADS-33 also does not specifically address the piloting requirement to provide a stable platform for the weapons system sights/sensors to be effective in acquiring, targeting and enabling firing on the enemy targets. With this requirement, the piloting often needs to be selectively handed over to some form of autopilot allowing the aircrew to concentrate on the mission aspects of the operation. Any opportunity to reduce pilot workload with regard to flight operations, effectively frees both the pilot and co-pilot gunner to focus more on the mission at hand, thus increasing the tactical effectiveness of the aircraft.
- DVE Operation – Certain advanced flight control features are required to make operating in DVE practical, particularly when the pilot loses situational awareness [Ref. 15, 16]. Obstacle avoidance cueing provides tactile cues in the affected control axes to the pilot, providing an intuitive natural interface that aids the pilot. Other features, including certain autonomy functions described later, when combined, provide a comprehensive DVE operations solution [Ref. 17]. In addition to the safety element of operating in DVE, substantial tactical advantages are gained from being able to operate an attack helicopter in such weather conditions. Not only does DVE operation allow for air support of ground operations, but also allows tactical engagement to occur when the opposing forces are at a tactical disadvantage. By interfacing with the mission systems equipment package as well as future enhancements such as improved radar, lidar and distributed aperture sensors, an enhanced flight control system would significantly increase safety and lethality of the AH-64 in DVE conditions.
- Improved Vehicular Performance – Studies have shown that pilots are not able to get the most out of the aircraft due to either concerns of exceeding vehicular limits or because limits are artificially reduced in order to ease pilot decision making [Ref. 18].
- Improved Mission Task Effectiveness – By coupling the flight control system to other tasks, the pilot can conduct those tasks more effectively. The studies show a 27% improvement in mission effectiveness when combining several metrics such as time to complete (19% improvement) and usable agility (35% improvement) [Ref. 19].
- Reduced Operating and Sustainment Costs – Studies have shown that advanced flight control features should significantly reduce helicopters O&S costs. For example, a study in 2015 showed the reduction in AH-64 O&S costs to be roughly $23,500 per aircraft per year (or 11.5 aircraft in a 20 year period), primarily due to reduced engine torque exceedances [Ref. 13]. The premise is that tactile cueing in the vertical axes (collective) dramatically reduces the likelihood of such exceedances.
- Improved Safety - Loss of helicopters due to tail rotor tree strikes, Controlled Flight Into Terrain (CFIT), power settling avoidance, and vehicle-to-vehicle contact have had a significant impact on cost, not only financial but also lives lost. The same above study has suggested a $53,000 per aircraft per year (or 26.7 aircraft over 20 years) savings [Ref. 13]. The automotive world has already implemented such technology with success.
- Autonomy – The term "autonomy" conjures up a variety of differing visions but it can range from various simple functions such as hold modes, which the Apache already has, to fully autonomous (no pilot) operation. The term Mission-Adaptive Autonomy (MAA) is used to describe that range and pioneering work has been done to develop and demonstrate key aspects of MAA, focusing on autonomy capabilities like obstacle field navigation, safe landing area determination, path generation and waypoint guidance [Ref. 20, 21]. In addition to those capabilities, many other autonomy capabilities would also reduce pilot workload, improve effectiveness, and improve safety for the future Apache. Example capabilities include automatic takeoff and landing, automatic autorotation, and autonomous/aided shipboard operations [Ref. 22]. Another focus for Boeing has been the Pilot-By-Exception/Selective Autonomy concepts whose purpose is to allow the pilot to easily and

intuitively enter into and exit from autonomous modes of operation.

- Coordinated Operations - There is current interest in cooperative operations between vehicles, sometimes referred to as "swarm operations." This could be either in the context of operating with other manned vehicles or small-unmanned vehicles, perhaps deployed from an AH-64, or a combination of the two. Such swarm operations has become de rigueur in some military circles [Ref. 23]. However, other very practical but mundane applications exist such as flying in formation for fuel savings and vehicular de-confliction for air traffic control.
- Higher Speed Operation – if the AH-64 were to ever be modified to support higher speed operation, for example by making it a compound vehicle (e.g. propulsor, other control surfaces), such a system would require a much more sophisticated control system. This control system would be required to map the controls to the surfaces under both hover, low speed and high-speed operation where the surfaces have different effectiveness. The implication is that additional control inputs, such as the introduction of additional effectors in the crew station such as a power lever or collective twist grip like that of the Cheyenne [Ref. 24], might be needed to accomplish this.

In addition to the new requirements described above, there are certain well-known existing requirements that need to be met and that play a large role in architectural decisions. Reliability is the primary example of an existing fundamental requirement for consideration in selecting a VMS Architecture. It is important to make the distinction between two different aspects of reliability: Safety of Flight Reliability and Mission Reliability.

- Safety of Flight Reliability - The most important reliability requirement is the basic safety requirement based on Probability of Loss of Control (PLOC). Although the numbers of key sensors and levels of autonomy affect the ability to determine or verify the PLOC for a given architecture, the PLOC number requirement is independent of the architecture or any of the performance or feature requirements mentioned above. However, with failure modes response, there is likely to be some variance in customer desires for the failure response as it is dependent on levels of autonomy or on the mission or mission phase. All candidate architectures will require the same Safety Reliability.
- Mission Reliability - The mission reliability is the Probability of Loss of Functionality (PLOF) that is required to complete the mission. Mission reliability can vary significantly based on customer desires and mission objectives. Example mission requirement features or functions range from the basic hold modes of the CLAWs, to haptic cueing providing obstacle avoidance cueing in DVE, to autonomous waypoint steering through tight enemy weapon detection corridors. One assumption is that once autonomous features are engaged and in control of the aircraft, that the PLOF for those functions needs to be the same as the PLOC. However, that may not always be the case if the pilot can take over when there is a critical failure in the automated system. If there is sufficient monitoring, an effective alerting system is in place and there is enough reaction time available for the pilot to safely take over, then the automated system does not have to have a reliability as high at the PLOC. This opens up opportunities for a much more affordable architecture that still provides capability with acceptable performance and reliability.

There are also other important implicit requirements such as minimizing data transport delay, providing deterministic, well synchronized data buses etc., but this study accepts such requirements as being needed to be met for all the proposed architectures and the paper does not dwell on them since each architecture addresses these requirements equally well.

CANDIDATE ARCHITECTURES

Four different candidate architectures for the AH-64 are discussed, as well as their various advantages and disadvantages along the stages of design, development, and production. The first architecture discussed, the conventional triplex FBW case, is probably the most familiar to most readers. The second architecture, APAS, may be familiar to some readers as it is emerging on the H-47. The last two architectures are specifically unique to the Apache because of its unique existing flight control systems design.

Traditional FBW

Traditional FBW is an obvious architectural candidate. It is well established that it can provide the most advanced control laws which improve piloting, simultaneously improving both maneuverability and platform stability, through advanced hold modes and response types. The improved platform stability allows the pilot to focus on mission tasks such as effective weapons deployment. Traditional FBW also readily satisfies the other requirements already discussed.

Figure 1 shows an example traditional FBW for the AH-64. The key architectural features of the traditional FBW architecture consist of the following elements.

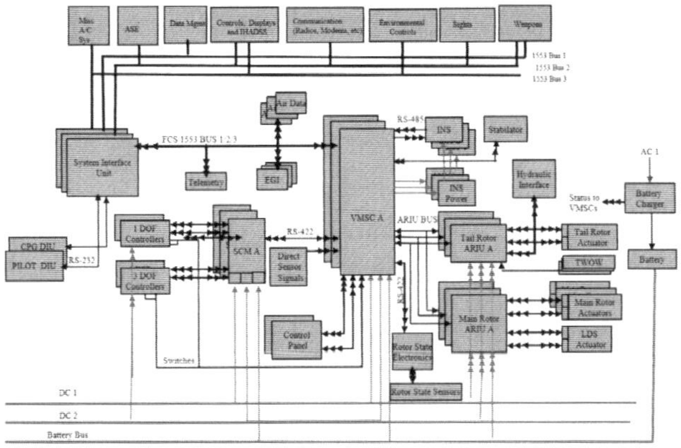

Figure 1: Conceptual Traditional FBW for an AH-64 System

Common (Similar) Triplex

In order to achieve the required safety and Probability of Loss of Control (PLOC) (while $1x10^{-9}$ for a commercial aircraft, military rotorcraft are more like $1x10^{-7}$ per SAE's AS94900), a triplex architecture is required for all components that provide safety of flight function. These include:

1) The control path between the pilot and the control effectors (actuators) that affect primary flight
   a. Control position sensors
   b. Signal paths between sensors and VMS computing
   c. Computing sources (VMS computers)
   d. Signal paths between computing and actuation
   e. Actuation to control surfaces
2) The aircraft state feedback control path between the state determining sensors and the control effectors (actuators) that affect primary flight
   a. State sensors such as INUs, GPS, radar altimeters, air data, etc.
   b. Signal paths between sensors and VMS computing
   c. Same as 1.c-e
3) Sensors that the pilot uses to determine the hazards to flight (other than pilot's eye such as in DVE)
   a. DVE sensors such as lidar, millimeter wave radar etc.
   b. Terrain or obstacle Database data sources
   c. Displays (HUD or cockpit)

In the case of traditional FBW, this triplex redundancy is accomplished through having three each of every component and also having actuators that accommodate three different input sources. This means that things like redundancy management, channel equalization etc. all need to handle all combinations of both single and dual failures which makes the software, and hence V&V, quite complex. In contrast, other architectures that we examine do not require this, albeit at the expense of feature or mission reliability.

Full Authority

Because the actuation is fully controlled by computer, not only is the pilot control path full authority, the automated control features such as stabilization and autopilot can be full authority as well. This is in contrast to traditional partial authority systems where the electronic command path to actuator that enables the flight control sophistication is limited (usually 10 to 20% of the full authority).

However, it is important to note that many implementations of FBW deliberately distinguish between the "primary" flight controls and the "automatic" flight controls, limiting the authority (either by position or rate) of the automatic flight control commands. In such cases, the primary flight controls are an extremely well vetted (tested) set of control laws with all the necessary redundancy that the designers have 100% confidence in. The automatic control laws tends to provide much more functionality, are much more complex, and rely on many more sensors that may not have the redundancy pedigree (either in quantity or testing). With this type of implementation, the FBW architecture's automatic control laws may suffer from some of the same performance limitations as partial authority systems.

Actuation

The new main rotor FBW actuators would be arranged on a new stationary swashplate where the pitch, roll and collective mechanical mixer is removed and replaced with electronic mixing provided by the software. This approach offers potential weight reduction. These main rotor actuators along, with the tail rotor actuators, would need to satisfy the redundancy requirements for FBW.

Developing FBW actuators has historically proven to be expensive and one of the biggest impacts to the development schedule. Fortunately for the AH-64, there have been several FBW programs for helicopters of similar size so new AH-64 FBW actuators would be able to leverage those programs. That said, a new actuator design would provide the opportunity for size, weight, power, reliability, and performance trades to accommodate potential future requirements. Potential future requirements that affect actuator design include a new higher performance rotor (such as a stiff rotor) or even supporting a higher speed (compound) helicopter design to include varying the Pitch/Roll control phasing in software that the legacy mechanical mixing does not allow.

Active Parallel Actuation System (APAS)

To take advantage of modern flight controls features such as trim management and inceptor based tactile cueing without the potential cost and complexity of a traditional FBW system, the concept of the Active Parallel Actuation System (APAS) system should be considered. Originally developed for the H-47 platform on the AVMS program [Ref. 8, 9], APAS adds electronically controlled active servo actuators in parallel to the mechanical controls that replace the mechanical force trim and feel system to provide an electronically controlled, dynamically changing force feel system. APAS provides both a programmable, dynamically changing trim position for the pilot controls throughout their full authority range which eliminates the issue of steady state SCAS saturation which is typical on most partial authority, mechanical systems. APAS also provides force feel (tactile cueing) features such as soft stops, notches and gates that can be used to cue the pilots to where the control positions need to be to prevent exceeding aircraft limits.

The legacy H-47 force feel system pallets consist of force feel springs, magnetic brakes and balance springs located in the flight controls closet. For the APAS modification, those pallets are replaced by electronic control feel pallets, which are arranged in pairs, where the pitch and roll are in one and the yaw and thrust/collective are in the other. Each electronic control feel pallet consists of a motor referred to as a Servo Actuation Unit (SAU) and a Force Sensor Assembly (FSA) for each axis. Figure 2 shows these basic components in the APAS context, albeit for a proposed AH-64 configuration. Between the SAU reduction gears and the FSA, there is a Clutch Friction Device (CFD) that engages when the system is operational and provides a level of friction that allows for some damping for the pilot controls when the clutch, and hence the APAS, is disengaged.

Another key component of the APAS system is the Force Actuation Control Electronics (FACE) which takes in commands from the flight control computer and feedback from the FSAs and controls the SAU which is mechanically linked to the cockpit controls. The FACE commands the SAUs to a position that corresponds to a force versus displacement curve that is programmed in the FACE software. The curve is intended to dynamically change as the control laws perform their function. The Force/Displacement curve is a generalized programmable electronic replacement of specific fixed mechanical force-displacement relationship that the legacy mechanical centering spring based trim system provided.

For the most basic APAS operation, the trim position is updated as the aircraft maneuvers and changes speed, heading or vertical speed during flight the same way a trim motor does in a traditional automatic or beep trim system. However, instead of the spring constant, mass, and damping being fixed as is the case for the legacy mechanical systems, those characteristics are programmable and dynamically changeable with the APAS. In addition, more advanced features such as softstops (position and force level) are also dynamically changeable as the flight condition warrants. During normal operation, the high frequency commands such as stabilization and command augmentation are commanded through the legacy SCAS system while only the trim commands and force feel characteristics go to the APAS.

A schematic of how the APAS system would fit into the AH-64 flight control system's architecture is shown in Figure 2. The concept for the Apache APAS implementation is to reuse the H-47 APAS's hardware components but not in the palletized arrangement. Because the AH-64 flight control elements are dispersed through the airframe and space is at a premium that limits the ability to provide a palletized solution, it makes sense to de-palletize the APAS components and distribute them. This also helps to minimize the modifications to the legacy platform.

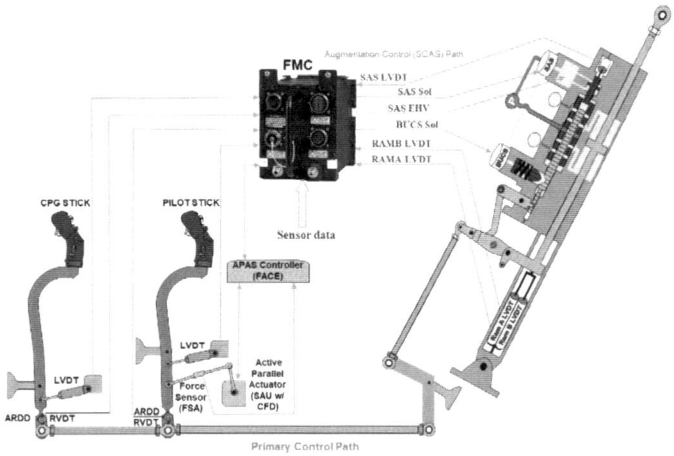

Figure 2: APAS on AH-64 Flight Control System

There are some other mechanical integration details to consider. The first one is whether to mass balance the control sticks to prevent the APAS from treating inertial forces measured by the FSA as pilot inputs when the aircraft is maneuvering. Mass balancing adds several pounds to the system but eliminates the need to compensate with software for the dynamic force imbalance. Another consideration is whether to replace the heavy clutches with a simpler decoupling mechanism to handle SAU gear train jams. Two options are a shear pin option similar to the ones used for the legacy trim spring and an option to use the AH-64's existing ARDDs that provide the pilots a means to break out of a mechanical jam and to enter BUCS mode. Both of these options should be explored, analyzed and compared.

The H-47 APAS computing architecture is a Dual channel, Single Fail-Operational system with each channel lining up with the H-47's existing dual redundant flight control computers. When the system detects a failure, the other channel takes over to mitigate failure transients. The FSAs are quad, two to each channels with each channel having a command path and a monitor path. This ensures a bad FSA measurement will shut down the affected channel but the other will be fully operational and able to remain active while still being able to detect a second FSA failure and gracefully shut down. The SAU motors have dual channel wiring. Each of the two H-47 pallets have a FACE computer which is dual channel for the two axes they control (either Pitch and Roll or Yaw and Thrust/Collective).

In contrast, for the AH-64, it is envisioned that the two FACEs be four axes for each FACE but the two FACEs be separate channels. This has several advantages including improved mission reliability and battle damage resilience. However, it is also important to note that, unlike the H-47, the AH-64's existing SCAS and flight control computer (referred to as the FMC) is single channel. A production AH-64 APAS would require the dualization of the flight control computer, but likely only to accommodate the APAS aspects and not the SCAS or the BUCS since doing so would be much more costly. Thus, the AH-64 APAS should consider the option of having both an FMC and another (smaller) FCC to handle the APAS as well as any other future features like autonomy.

To satisfy increased flight control system and mission reliability requirements likely for an Advanced Apache, redundancy in the automatic flight control path will be required so that the system will retain trimming, stabilization and autonomous features after a failure. Also, the APAS's SAU control has the bandwidth to provide stabilization inputs if the single channel SCAS system fails. This would differ from the H-47 APAS which does not do this because its SCAS is dual channel. However, in order to use the APAS in this way, further software certifications would need to be addressed since the system is currently only certified to provide rate limited position commands based on FSA force measurements and adhering to the Force Feel model curve. There are several examples of parallel actuator stabilization systems using this concept for helicopters, including HeliSAS and Helitrack, so this should be an option to consider.

The primary advantage of the APAS is that it would seem to achieve most of the features that the FBW system has without incurring the complexity and cost of replacing and certifying new actuators and flight management computers as well as the high cost of the software certification to support a triplex architecture. The next section examines the extent we feel that it can achieve the various features. Another advantage is that the APAS hardware from the H-47 APAS program can also be largely re-used.

There are some challenges when trying to implement APAS on the AH-64. The inherent APAS redundancy is Dual, so to get the reliability needed for autonomous operation, there will be a need for a non-standard dissimilar flight control computer architecture.

Another challenge is that the existing SCAS authority is only 10% which may make it difficult to take full advantage of more augmented response types such as TRC and ACAH. The existing actuators' authority can be readily increased to 20% but doing so requires either the ability to show that pilots can recover from a 20% hardover or that the system can detect such failures and disengage in a timely fashion. Another way to get these more augmented response types is by dynamically moving the inceptor trim position but that can produce undesirable control forces which may confuse the pilot.

Finally, the most notable challenge for the APAS is to ensure that the APAS does not interfere with or otherwise affect the Apache's Back-Up Control System (BUCS) operation. The BUCS is a system where each of the four main hydraulic actuators have a FBW mode that is activated when a flight control linkage severance or jam is detected. For the jam condition, depending on where controls are jammed, there are decoupling devices that are designed to decouple with a specified amount of force and displacement (which are defined differently for each axis and for each of the two pilot stations). The decoupling devices, referred to as ARDDs, also have RDVT sensors to detect when the decoupling takes place. For the control severance condition, there are mistrack algorithms, which monitor the differences between the control and actuator positions, in the software to detect a severance. These mistrack and decoupling signals are used in the FMC to command the actuator to operate in the BUCS mode. When the actuator is in the BUCS mode, the mechanical input to the actuator is locked-out so that the actuator responds only to electronic inputs, making the system a single strand FBW system.

Because the APAS applies forces to the controls, it can contribute to, or interfere with, the forces that cause the decoupling devices to decouple. If the APAS forces were identical to the legacy mechanical force feel system, then that would not be an issue. However, variable softstops, damping and inertia are what provides APAS value and those provide different forces than the legacy mechanical trim system. Therefore, testing of all the BUCS engagement scenarios would need to be performed in a representative environment to ensure the APAS does not negatively change the behavior of the BUCS operation. Further, BUCS was only meant to be engaged very rarely and, as such has only been engaged in a very limited amount of times over the course of the AH-64's history. Using the ARDDs as a means to handle an APAS jam would increase that frequency and the effects of increasing that frequency would need to be analyzed. Furthermore, the BUCS engagement logic is quite complex and it needs to be verified that using the ARDDs as a means for APAS disengagement does not affect that logic. Finally, there is also the need to consider the interaction of APAS with gunfire, particularly in regards to the FSAs' performance and their force feedback into the FACE. The effects of gunfire on APAS is difficult to predict or quantify without actual gunfire time in flight.

APAS with Coupler

The baseline APAS has two key challenges: how it manages trim and how it commands higher augmented response types such as TRC and ACAH. With the baseline APAS, the cockpit controls are never separated from the surface control actuators by more than the amount of SCAS authority and are aligned when in steady state to allow the full SCAS authority to provide stabilization. The ADS-33 HQ specification states that the more augmented response types such as TRC and ACAH are required to achieve Level 1 HQ in DVE but those response types require mechanical separation between the cockpit controls and the actuators. This separation can be achieved by adding a decoupler between the two. Decoupling the controls also results in less movement in the cockpit controls since the cockpit controls are no longer forced to follow the actuator motion. There are more details of this concept in the Architecture Implications on the Requirement's Improved Handling Qualities and Pilot Workload subsection.

For these reasons, adding a coupler to the APAS architecture provides another architectural candidate. The original coupler concept was developed in 2007-2008 time frame as a means to provide an inexpensive FBW system for the AH-64 [Ref. 12]. The concept was to nominally operate in BUCS mode with the coupler in the decoupled state. A failure of the single strand system (albeit with redundant sensors) would then result in re-coupling the two stations to the mechanical control linkage so that the pilot could then fly the aircraft through the mechanical flight control system. This provides a single strand FBW system, with the benefits of the advanced control laws offered by FBW, but without the cost of triplex redundancy. It requires excellent robust fault detection, but also needs to account for servo actuator command transients (along with vehicular motion transients) that would ensue as the system recoupled. It also enables a largely single strand (simplex) FBW system with other components (such as inertial sensors) that would typically require triplex redundancy for use in a FBW system, only needing to be duplex in order to be single fail operational and dual fail-safe.

Adding APAS to the mix allows tactile cueing in FBW mode, but also when mechanically recoupled. Figure 3 illustrates the APAS System with Coupler concept. The only difference between it and the original 2007-2008 concept is the addition of the APAS.

The original system had two decoupler designs. The first one is a limited authority spring decoupler. When in the primary (FBW) mode, this spring continuously provides a force proportional to the displacement the cockpit inceptor is from the surface actuator such that when the pilot force is removed, they are aligned (minus the SCAS series actuator input). The second design has no spring force during normal operation and decouples the full range of control motion.

Both decoupler designs have challenges in the transient that occurs when reverting from FBW to mechanical mode. When the FBW mode is failed and the actuator reverts to mechanical mode, the force applied to/by the pure spring decoupler will move the actuator to the cockpit control position thereby causing a transient. When reversion takes place with the alternative decoupler, a damped spring is engaged to convert the decoupler to a rigid link. Since the spring is more damped, the transient is less abrupt than the pure spring decoupler as it extends or retracts to its rigid length. The APAS can be used to reduce or eliminate these transients by providing the necessary transient force and stick position to keep the actuator from its transitory movement during the reversion and make aircraft transients indiscernible. However, keep in mind that when the FBW stick to actuator relationship changes from FBW to mechanical mode, an eventual stick position change is required to prevent an actuator transient.

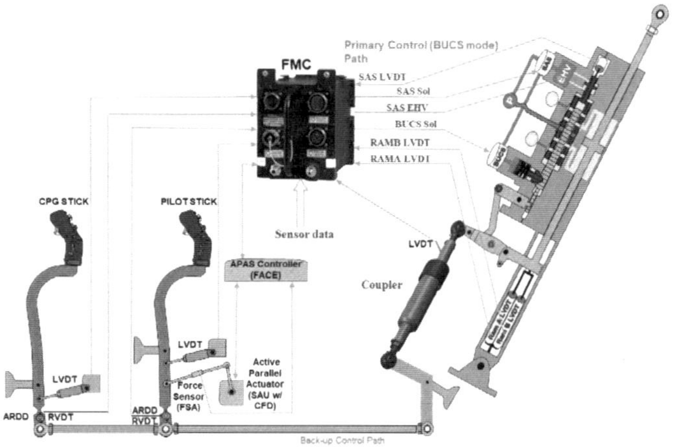

Figure 3: APAS System with Coupler

Combining the coupler and using the FBW mode as the primary flight control mode with the APAS architecture retains all the advantages of APAS, including control strategies and tactile cueing, while the added authority for the inceptor to move independently from the control surfaces to enable better handling qualities. Adding the coupler, which opens up the inceptor authority independence from the actuator, also provides more potential for compound flight control strategies where the inceptors may need to move multiple control surfaces with differing stick gradients and or feel characteristics.

On the challenge side, adding the decoupler and using the actuator's FBW mode as primary control adds both mechanical and software complexity that will require a more robust signal and failure monitoring system than the legacy or baseline APAS system. The mechanical complexity may increase the maintenance costs due to more involved rigging procedures and more failure modes that need to be considered.

Dissimilar Redundancy Extensible FBW

Another architecture option is what we refer to as Dissimilar Redundancy Extensible FBW. The feaures of this architecture is that it provides several distinct stages allowing for growth opportunity and flexibility for the customer to choose the capabilities it can afford as program priorities mature over time. This dissimilar FBW architecture concept can gradually mature to achieve higher levels of redundancy and mission reliability as budgets allow. The first stage, shown in Figure 4, while the most elementary and least costly, still provides full FBW capability and acceptable PLOC reliability. Increasingly improved levels of reliability can be implemented in stages, with a pre-production first stage already being demonstrated on Apache as part of the AVMS program [Ref. 25].

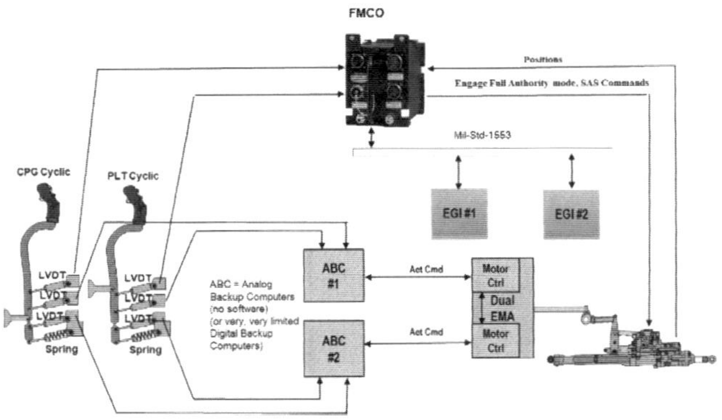

Figure 4: Analog Backup Computer (ABC) Stage Dissimilar Redundancy Extensible FBW

The key characteristic of this FBW design concept is to use the BUCS single strand FBW mode of the Apache actuators as one of the redundancy control paths. Two additional redundancy control paths, which are required to achieve acceptable PLOC reliability for a FBW system without a mechanical backup, are achieved by adding two low power actuators to the servo's mechanical input. Those actuators, because of their low power/force requirements can be very small electro-mechanical devices. Further, in order to reduce development cost, these actuators would be controlled with very simple computers that are either analog (op-amp based) or have minimal software to provide a direct electrical link between the pilot controls and the main actuator if the primary single redundant (BUCS) path fails. Since no or very little software would reside in the dual EMA and the computers controlling it, certification should be much easier to obtain than the traditional FBW approach.

The first stage has the following key design elements:

- Primary single channel full fly-by-wire via BUCS, with dual channel fail-operative direct electrical link.
  1) Dual direct electrical link designed to be at least as robust as the AH-64's mechanical controls
  2) Guarantees reversion to electrical EMA command path when needed
- Robust & comprehensive fault detection, isolation, and reversion.
  1) For sensors, actuators, and FMC
  2) Fail-operative control system modes for all failures that can be detected and isolated and have redundancy.
- Actuator monitoring extensive built in test capability

This initial architecture can be gradually matured to provide more capability, mission reliability and better failure transient response.

The next stage, a much more robust architecture from a mission and advanced functions reliability perspective, is shown in Figure 5. The figure shows dual extensible Flight Control Computers (eFCCs) controlling the EMAs. The eFCCs would have software that provides the redundancy management, signal management and control laws similar to that of a traditional FBW system as a part of a dissimilar triplex FBW architecture.

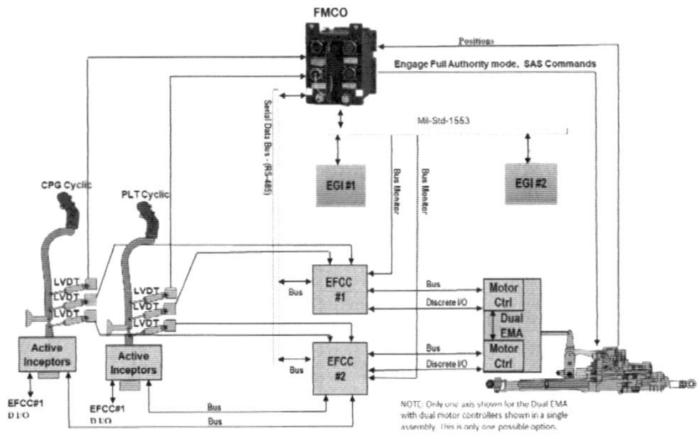

Figure 5: EFCC Capable Dissimilar Redundancy Extensible FBW

The EMA paths would be used as the primary control paths in conjunction with the partial authority SCAS control path. This approach mitigates the challenges of having a single FMCO self-detect internal failures (though it currently judged adequately in doing so for the production AH-64) and it also reduces failure transients compared to what could be experienced when using the single channel BUCS mode as primary. This is because the second EMA path would reject (fight) the failed signal command until the competing channels can be turned off as the BUCS path takes over. In contrast, for the single channel BUCS mode, the FMCO would need to monitor the main actuator operation and shut off the BUCS mode if it sees the actuator not tracking where it was commanded. This failure detection and response is not instantaneous so the effects of transients needs to be analyzed and, if unacceptable, mitigated.

Various components such as active inceptors and sensors can be added to improve both functionality and mission reliability based on funding available and requirements needed to be met. Figure 5 includes the active inceptors but not additional sensors. Figure 6, a representative last stage, shows the addition of redundant MEMS IMU state sensors. This last stage provides the capability closest to the Traditional FBW architecture.

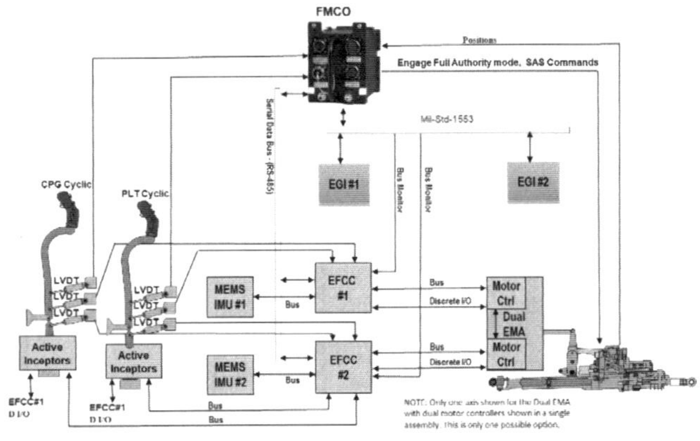

Figure 6: MEMS IMU Dissimilar Redundancy Extensible FBW

In summary, the extensible, dissimilar architecture provides flexibility to achieve the desired FBW capability that the customer chooses to afford based on various funding profiles and cycles. It also provides an architecture that can gradually mature with additional incremental hardware and software modifications to improve reliability, performance and capability. The architecture extensibly supports the full FBW capability to support complete autonomous flight and future compound flight controls if desired. The architecture reuses the most costly to design and certify components such as the FMCO and the main servo actuators and, therefore, is a quicker path to production than the Traditional FBW architecture.

Other Architecture Options

While presenting four architectures, it is important to emphasize that it is possible to form new architectures by selecting aspects from the various previously discussed architectures to provide the desired capabilities within the customer's available budget. The idea is to let the required capabilities and their priorities drive the flight control system improvements to be implemented. The following is an example path for the progression of increasing flight control system capability.

First, the most simple of an APAS installation, could be performed. This simplest installation would involve only replacing the mechanical feel spring with APAS hardware and minimally modifying the FMCO software. This would keep the existing control laws but modify them to update the trim position to eliminate steady state SCAS saturation. Then, software modifications that increase the capability by revising the control laws to provide advanced response types and tactile cueing can be made as they are budgeted for. The computing hardware modifications to implement APAS could be made with a computer such as the eFCCs in the Dissimilar Redundancy Extensible FBW Architecture with the I/O spare capacity to support potential growth to an extensible, dissimilar redundancy extensible FBW architecture solution when funding becomes available. Though the computing hardware supports growth for a dissimilar FBW, once APAS is in place it can also easily support the APAS+Coupler architecture if that path is desired.

However, the APAS+Coupler may not support, or at least as well, all of the customer's requirements. For example, if the autonomous capability becomes a higher priority to accompany the improved handling qualities requirements, the extensible architecture (IV), which provides improved reliability over APAS+Coupler, can be implemented leveraging the APAS architecture and the software using the actuator's BUCS mode as a redundant control path that are already in place.

ARCHITECTURE IMPLICATIONS ON THE REQUIREMENTS

Previously, we had been discussing the various architectures that would allow FBW-like capability to be brought into realization on the AH-64. While there was some initial discussion on pros and cons of each of the architectures, this section provides a much more detailed discussion about how well each of the candidate architectures accommodates the various requirements. For convenience and conciseness, we assign a roman numeral to each architecture, using the roman numerals as a reference tag in our discussions.

I. Traditional Fly-By-Wire (FBW)
II. Active Parallel Actuation System (APAS)
III. APAS with Coupler
IV. Dissimilar Redundancy Extensible FBW

In addition to qualitatively discussing the ability of the candidate architectures ability to support the various requirements, we also assign a rough measure of value for each architecture's ability to accommodate that requirement. We use a scale of 1 to 3 because that is an appropriate resolution based on this being only a preliminary analysis. Within the 1 to 3 scale we use a "+" or "-" to capture what we believe are meaningful differences between architectures. The values assigned should be used in the context of trends (i.e., is one architecture clearly better or worse than the other), as well as an indication of where the authors' concerns lie. Future work should refine these scores with more rigorous analysis, simulations, and data. For the fourth architecture, we score it for each of the three major phases described.

Improved Handling Qualities/Reduced Workload

Improved HQ/RW is fundamentally addressed by the application of control laws that have advanced moding and response types, using regime recognition to automate the moding, and the introduction of tactile cueing features. Advanced control laws with these features can only be accomplished with full authority systems. While all the candidate architectures are full authority, we also need examine whether their design is conducive to the various control laws features. In this section, we discuss the extent to which each of the candidate architectures impact the advanced control law features.

Response Types

First, advanced control laws require more augmented control responses beyond the Rate-Command (RC) and Rate-Command Attitude Hold (RCAH.) These less augmented response types are more amenable to being implemented in mechanical with partial authority SCAS rotorcraft systems such as the AH-64, H-47, and UH-60. This is because rotorcraft naturally have a near RC response. For example, the current production AH-64 implements a (quasi) Rate-Command mode. RCAH is easily realizable on the AH-64 with engagement of the attitude hold mode. As the AH-64's current production flight control system is mostly +/-10% SCAS partial authority, RC and RCAH mode are the only control responses that can be readily implemented, though higher augmented models like TRC can be implemented in a small envelope.

The more augmented control responses to improve handling qualities and reduce workload, are ACAH (Acceleration Command Attitude Hold), ACVH (Acceleration Command Velocity Hold), ACPH (Acceleration Command Position Hold), and TRC. These response types are discussed in detail in documents such as the Background Information and User's Guide (BIUG) [Ref. 26]. We discuss each of these here so that they can be readily related to the architecture candidates.

The BIUG's formalized definition of the Attitude Command response is that a step command in inceptor pitch/roll controller force/deflection input will produce a proportional pitch/roll attitude change within 6 seconds with the resulting ground-referenced longitudinal or lateral acceleration being constant or trending asymptotically towards a constant if the final attitude is not constant. It is important to recognize that ACAH requires the ability to manually re-trim so that the operator can zero out inceptor forces at any achievable steady attitude. The BIUG mentions that combining a trim follow-up system, such as must be the case for Architecture II: APAS, can lead to objectionable handling qualities when operating in ACAH in DVE or with divided attention operations. This potential issue warrants more study for Architecture II if an ACAH response type is required.

ACVH is a response type in which stick displacement commands a vehicular acceleration from a trim velocity set point. When the stick or inceptor is returned to neutral/detent the vehicle holds the speed. This can also usually be combined with Flight Path-vector Command (FPC) on the vertical axis so that stick displacements or forces on the collective input result in holding the flight path angle that is maintained even if the vehicle is sped up or slowed down. While the authors don not believe that ACVH should have the same concerns as ACAH in DVE for the APAS architecture, this still needs to be verified.

We now proceed to examine the near hover regime control responses. TRC is a control response type where the pitch and roll controller inputs produce a proportional steady translational rate with respect to the earth (ground speed) in the longitudinal or lateral direction. This control response is useful when attempting to obtain Level 1 handling qualities for certain mission task elements such as fine hover repositioning when external visual cueing is extremely poor. TRC is a challenge in APAS because it can only be implemented in a limited envelope unless automatic stick retrimming is done. Automatic stick retrimming is a challenge from a pilot acceptability perspective and brings up the same issues commented on for ACAH in DVE.

Position Hold and Heading Hold are other important response types for DVE and divided attention but also for weapons deployment. While our experiences have led us to believe that position hold works well for all architectures, heading hold has been an issue for APAS. The issue lies in the fact that the directional (yaw) axes has significant coupling on a single main rotor helicopter. This coupling, which results in pedal back-drive (see ensuing Decoupling and Pedal Back-Drive discussions), became apparent in AH-64 flight simulator evaluations of APAS. A couple examples of flight conditions where this phenomenon showed up were the Accel-Decel MTE and turn coordination during up-and-away cruise flight. Maneuvers such as these result in the pedal trim position changing substantially which is a problem because the pilot is not able to deal with that well, their feet not being sensitive or reactionary enough to let the pedals backdrive. While, a couple of promising methods have been developed to address this (per the Pedal Backdrive discussion) , these design solutions still have a tendency to interfere with the heading hold performance. Future work is needed to come up with a solution that best addresses these competing issues. In summary, improved handling qualities and reduced workload has been shown to require advanced control responses, all of which are readily mechanized in all the architectures except APAS.

Unique/Non-Unique Trim

Another related aspect to the trim follow-up (e.g. non unique-trim) discussion is that some aircraft/control law designs lean toward unique-trim and others toward non-unique trim. The non-unique trim provides the advantage that the pilot does not lose awareness of the current rotorcraft control surface state. In this regards, the FBW architectures (I and IV) are the most flexible since they can easily be reconfigured between unique trim and non-unique trim on the fly, whereas architecture II (the pure APAS configuration) only supports a non-unique trim system. Architecture III can support unique trim for the full-authority coupler but not the limited authority coupler, and even the full authority will have to face bigger transients during recoupling if implementing unique trim. Architecture III can support unique trim for the full-authority coupler but not the limited authority coupler, and even the full authority will have to face bigger transients during recoupling if implementing unique trim.

Regime Recognition

Regime Recognition is a technique where sensor inputs are used to determine the flight condition the vehicle is in with respect to what kind of control strategy is best for that condition. In some ways, it can be thought of as a unified state estimation process that generates information that the control laws act upon. Examples of regimes that are typically recognized for their unique control strategies include Hover, Low Speed Cruise, Cruise, Flight Path Vector, Autorotation, and Ground Mode. In addition, there are differences between turn coordination at lower speeds (nose to tail trim at speeds between 20 and 60 knots) and at higher speeds (ball-centered trim at speeds > 60 knots). Regime recognition is primarily based on sensed aircraft states that are fed to the flight computer. For the no-failures condition, the sensed aircraft states will be the same for all four architectures. However, for the case of a failed aircraft state, the triplex sensor systems (Arch I, IV.3) will be able to resolve this and continue as normal. However, the non-triplex systems will not necessarily be able to fault isolate. Provided the system is still able to operate with the core control laws (which may not be the case depending on the state failed), the system will have to disengage those regime recognition features in the control laws that need to use that state for which the sensor failed.

Tactile Cueing

Tactile Cueing is a method to enforce envelope protection of the vehicle by giving the operator a feel for where the limit is, and requiring them to either consciously push or pull through this boundary if they wish to exceed the limit. Some good examples of collective axis tactile cueing limits include engine torque, engine gas generator speed, rotor speed, engine temperature, and descent rate limits. These cues have been demonstrated during the AVMS program with Architecture II on an H-47 [Ref. 9] and with a pre IV.1 stage architecture on an AH-64 [Ref. 25]. The results from those studies showed tactile cueing to work well on both architectures and it is expected to work well on the other architectures as well.

Decoupling

Decoupling is the process in which the Multiple-Input Multiple Output (MIMO) control problem for a vehicle control is transformed to approximately a Single-Input-Single Output (SISO) system. For a single MR-TR full-authority helicopter, this can be accomplished by use of an invertible 4x4 decoupling matrix that converts the ideal control vector (body angular and vertical accelerations) to the actual realizable control vector (delta actuator positions). For full-authority FBW architectures (e.g. I, III, and IV) decoupling can be implemented in a relatively straightforward way by inverting a 4x4 matrix either sequentially or all-at-once. However, with the APAS architectures (II), the decoupling becomes more complex. The complexity arises from the fact that APAS, since it still retains the mechanical path between inceptor and actuator and has an authority limited SCAS, needs to use the SCAS for high bandwidth/high rate commands, and needs to move the stick for low-rate or steady state commands. The decoupling function in the control laws needs to properly apportion this for all conditions and maneuvers. This decoupling complexity for Architecture II, amplified by single main rotor helicopters having significant inter-axis coupling, results in it rating a little lower for Architecture II.

Pedal Back-Drive

The inceptor trim command that results from the decoupling approach for Architecture II results in the backdriving of the inceptors to change the inceptors' trim position. Depending on the speed and extent of the trim position change, the pilot can find this annoying and, in some cases, problematic.

This is particularly true in the pedals for a single main rotor helicopter (in contrast to a vehicle like a tandem, coaxial or multicopter where directional anti-torque not an issue). For a single main rotor helicopter with mechanical controls, a significant amount of pedal motion is required to counteract the multi-axes aerodynamic coupling. This is amplified because of the friction interaction between the operator's heels and the floorboard and the fact that the tactile response of human feet are much different than human hands. Although a non-unique trim backdriving inceptor in an operator's hands may be a nuisance, back-driving pedals can result in serious unwanted control inputs and flight performance or safety issues.

A couple of promising methods have been developed to address this, including a variable rate damper control law architecture [Ref. 27], and a variable operator interface [Ref. 28]. Unfortunately, per the Response Type discussion, these design solutions still have a tendency to interfere with the heading hold performance. Future work is need to come up with a solution that best addresses these competing issues.

The conclusion is that the various full authority FBW architectures (I, III, and IV) don't have issues with pedal back-drive whereas Architecture II has challenges.

Summary

Table 1 aggregates the various control law discussion into a single assessment for each candidate architecture.

Table 1. Scoring of Candidate VMS Architectures with respect to HQ/WL

	Candidate					
Criteria	I	II	III	IV.1	IV.2	IV.3
HQ/WL	***	**+	***-	**	***	***

DVE Operation

DVE operation increases flight criticality of many of the sensors and automatic FCS components. That implies a requirement for additional reliability and hence redundancy which drives our architectural requirements. Autonomy also drives reliability requirements similarly as DVE operation drives the pilot to rely on more autonomous features of the

FCS. Therefore, both DVE and Autonomy need to be a significant part of the reliability discussion. Unlike autonomy where there may be a calculated decision by the customer to risk tolerance for losing an unmanned aircraft, for DVE capability it is assumed to be in the same context of safety of flight for crewed operation.

The DVE reliability discussion has the following elements:

1) A sensor becomes flight critical if the pilot needs it to safely operate in DVE. That flight criticality implies the need for redundant sensors.
2) Although a sensor might be flight critical while in DVE, one might think that it would be okay to live with mission reliability requirements because the pilot could simply abort the mission and leave the DVE environment by flying above or below the obscurant. However, in practice this might either not be possible because of where the aircraft is in the mission and the risk. Thus, having only mission reliability requirements might not be advisable.
3) If mission reliability is deemed satisfactory, then, depending on the reliability of the sensor, the system could seemingly be either simplex or duplex. However, this is further complicated by the reliability of detecting a sensor failure and how the sensor data is used (e.g. for autonomous functions).

A flight control system with integrated DVE flight control features (e.g. tactile cueing obstacle avoidance, formation flying pilot-by-exception) has similar reliability concerns. That is, the flight control system will need to be sufficiently redundant and with adequate failure detection monitoring in place unless the customer is willing to accept the lower reliability and associated risk. Mission reliability is much different for DVE than it is for most other functions as the loss of DVE capability is likely to imperil the aircraft more than those other functions. A possible scenario is where the crew operates with no system failures and can trust the DVE automated system with little to no supervision. However, after a single failure (of a duplex system), the crew still uses the automated features such as obstacle warning, waypoint navigation etc., but has to use them with caution or, not fully trusting them, exiting DVE by climbing to a safer altitude, or mitigating risk by slowing down to improve reaction time.

With these considerations in mind, Architecture I has the best ability to satisfy DVE since the triplex architecture with sensor failure mitigation is already fully embedded into its redundancy management scheme. Architectures IV.2 and IV.3 have nearly the same capability with the small exception that some failures will result in reversion to an actuation path that, because of its lower bandwidth or rate, might be less ideal than the single channel failure case of Architecture I. One example is losing the SCAS/BUCS channel where all the (trim and stabilization) control would have to come from the EMA path alone or switching to/from the actuator BUCS mode. Architecture II has lower capability for DVE operations because is it has to manage trim and challenges associated with providing TRC and ACAH response types known to be useful in DVE. Architecture III has better DVE capability than II due to the decoupling of the cockpit controls to the surface actuators. Both II and III score a little better than IV.1, the least capable for DVE, because at least they both provide tactile cueing where IV.1 does not. Table 2 summarizes.

Table 2. Scoring of Candidate VMS Architectures with respect to DVE Operations

	Candidate					
Criteria	I	II	III	IV.1	IV.2	IV.3
DVE Op	***	**	**+	**-	***-	***-

Autonomy

The main purpose of implementing autonomy on the AH-64 is to make it a more effective attack platform while making the vehicle safer and more efficient. By passing pre-determined, well-defined tasks over to an autonomous system, the operator is able to attend to more mission critical tasks. Some AH-64 relevant examples of autonomy include the following functions:

- Automatic Takeoff and Landing
- Autonomous Taxi
- Automatic Autorotation
- Automatic Weapons Aiming and Firing (also known as Integrated Fire & Flight Control)
- Automatic Flight Path Guidance (Waypoints, Trajectory Generation, and Routing)
- Automatic Shipboard Landing
- Selective Autonomy

While most of these autonomy functions are self-explanatory, one exception is "Selective Autonomy" which we now define. Selective autonomy refers to any sort of partial autonomy function that can be easily and intuitively turned on and off, keeping the operator continually in the loop to perform flying operations if needed, but maintaining the desired autonomous feature when the pilot does not want to be in the loop. This is in contrast to the traditionally viewed autonomous operation where the pilot has to perform an explicit handover of complete vehicle control to the flight control system with the operator then needing to go through some explicit handover to take back control.

In terms of which architectures favor the various autonomy functions, the answer is based on the reliability and capability limitation implications of those architectures on each of the autonomy functions. With respect to reliability, once a certain function is automated, the reliability must be high enough that the operator's judgment does not lead the operator to perform the function themselves, albeit at a higher workload. That means that, the operator must have very high confidence that the autonomy function will perform reliably repeatedly. Furthermore, if, for some reason, the autonomy function cannot be performed reliably (e.g., due to failure), the operator needs to know this as soon as possible so they can re-take control of the vehicle. In terms of capability, the autonomous function must perform well enough that the operator is willing to let the computer perform it with the knowledge that operator intervention may be required to take over. A prime example of capability limitations is that certain architecture's authority limitations will impact the ability of that architecture to perform that autonomous function well enough to be deemed acceptable by the pilot. There is an interesting phenomenon here because in order to get the performance desired, the autonomous function may require full authority, but this opens up the system to flight control system failures with larger transients where the operator might be less trusting of the autonomous system because these failures transients could lead to an unrecoverable state.

All of the autonomy functions listed above are subject to these previously discussed considerations. How each of the architectures rate against the ability to satisfy the autonomy requirements aligns relatively closely to the DVE requirements because the autonomy considerations are somewhat similar to the DVE considerations. Both put a high premium on the high degree of reliability and high performance/capability. One difference is that autonomy is a little more sensitive to reliability and performance shortcomings (given that the autonomy is a broad area covering both crewed and uncrewed operation whereas the DVE operation is assumed to be crewed). This manifests itself particularly in IV.1 which has only one truly FBW channel with the other two channels being direct electrical links. Such a system is far from ideal for the reliability required of autonomy functions. Table 3 summarizes.

Table 3. Scoring of Candidate VMS Architectures with respect to Autonomy

	Candidate					
Criteria	I	II	III	IV.1	IV.2	IV.3
Autonomy	***	**	**+	*+	**+	***-

Vehicular Performance, Task Effectiveness, O&S Costs, and Safety

The requirements section discussed the types of flight control features that result in improved vehicular performance, improved mission task effectiveness, reduced O&S costs, and increased safety. All of those functions rely on either advanced response types, advanced modes, and tactile cueing. Thus, the ability of each of the architectures to provide them is the same as their ability to support the control laws discussed in the Handling Qualities & Pilot Workload discussion. The result is that architectures I, IV.1 and IV.2 readily provide all of these features. Architecture II, III and IV.1 suffer from the same issues brought up in the Handling Qualities & Pilot Workload section. The results are summarized in Table 4.

Table 4. Scoring of Candidate VMS Architectures with respect to Performance, Task Effectiveness, O&S Cost and Safety Improvements

	Candidate					
Criteria	I	II	III	IV.1	IV.2	IV.3
Performance	***	**+	***-	**	***	***
Task Eff.	***	**+	***-	**	***	***
O&S Costs	***	**+	***-	**	***	***
Safety	***	**+	***-	**	***	***

Coordinated Operations

While Coordinated Vehicle Operations is in many regards similar to Autonomy, we distinguish between the two because Coordinated Vehicle Operations requires explicit cooperation between vehicles. The ability to adequately and reliably cooperate, and the resulting vehicular motions based on the coordinated operation, might be affected by choice of architecture. Two broad areas of coordinated operation are considered from an architectural point of view: coordinated formation flying and coordinated swarming (Swarm Autonomy).

<u>Coordinated Formation Flying</u>

Currently, the extent to which rotorcraft fly in formation varies significantly, doing so more for transporting themselves en masse from location to location, but not for prosecution of a mission where flying in formation puts them at additional risk. However, with the desire to fly in DVE, there is expected to be a greater need for flying in DVE for portions of prosecuting a mission. In addition, , studies have shown that flying in formation with a particular spacing can also provide significant fuel savings. However, the workload and risk of closely flying in formation and with particular spacing would be high without some form of automated coordination to replace the human coordination and piloted tasks. Hence, the Formation Flying Pilot-By-Exception (FF PBE) concept which allows a lead aircraft to be flown by the pilot (or autopilot) and the other trailing aircraft automatically follow in unison via a datalink between aircraft communicating aircraft state information). The system is designed using tactile cueing to allow the pilot to intuitively adjust relative position or breakoff altogether [Ref. 19].

Three architectural features are required to support the FF PBE: 1) the ability to adequately control the control surfaces for all anticipated maneuvers, 2) the need to only be fail-safe (vs fail-operational) and 3) the ability to backdrive the sticks for intuitive pilot interaction. The FF PBE feature is really only meant for non-aggressive maneuvering flight so all architectures, including Architecture II which has limitations with SCAS authority and with the speed at which the system can retrim, should be equally acceptable. Further, because the FF PBE is for piloted operation that only needs to be fail-safe (rather than fail-operational), it is also supported by Architecture II. Finally, all the architectures support backdriving the controls. Thus, all architectures support this concept well.

Swarm Autonomy

Swarm autonomy and associated multi-robot system control law configurations is starting to take traction in military aviation operations [Ref. 29]. For example, [Ref. 30] and [Ref. 31] indicate that part of the US Army Aviation's future roadmap involves exploring the possibility of integrating ALEs (Air-Launched Effects) on both legacy and future attack platforms including the AH-64. While some ALE swarms vehicle will not return to the aircraft, others will (e.g., scouts, decoys). To be able to interact safely with these swarms, the AH-64 state information will need to be highly trusted and highly reliable and the vehicular motion will similarly need to be very accurately and reliably controlled. Options to go fail-passive are not acceptable without significant consequences. This last point points to a problem with the APAS, APAS+Coupler and the initial Dissimilar Redundancy architectures compared to the other architectures. Table 5 shows the combined rating of both types of coordinated operations discussed.

Table 5. Scoring of Candidate VMS Architectures with respect to Coordinated Operations

	Candidate					
Criteria	I	II	III	IV.1	IV.2	IV.3
Coordinated Operations	***	**	**+	*+	**+	***-

High Speed Operations

Consideration for high-speed operations is under the premise that a thrust device (rear propeller) is part of the future Apache control effectors and that the control strategies for high speed will be different from that of a traditional helicopter. Specifically, for a single main rotor compound helicopter with a thruster, the collective pitch requirement goes down with increasing airspeed as the thruster takes over forward speed control. This effect is even more dramatic with a wing used to offload rotor lift. Along with this, high-speed flight control strategies mimic airplane type control such that cyclic inputs more directly affect vertical rate as power inputs affect speed. An additional cockpit thrust controller is needed to control thrust of the rear propeller. Most importantly of all, the airplane and helicopter control strategies do not align with each other and to have the same cockpit inceptors for both strategies, the inceptors need to be physically separated from the helicopter control surfaces.

With these considerations, some VMS architectures are more amenable to satisfying these requirements than others. Arch I (and later stages of IV) are clearly the most amenable to mitigating this issue since they are fully FBW and would support all control strategies throughout the entire speed envelope for single and dual failures. Architecture III with the full authority coupler also supports compound operation before any failures but would need to revert to non-standard control strategies to continue high-speed flight after a FBW actuation control failure. Arch III with the reduced authority spring coupler would have limited capability to provide compound operation as long as APAS is used to counter the coupler spring force needed to have dissimilar inceptor and swashplate commands. Similarly, after a FBW failure, recoupling to mechanical FCS would entail non-standard control strategies to allow continued high-speed flight. Architecture II is the least amenable architecture for compound operations due to the mechanical linkage to the main (helicopter) control surfaces as it would always require non-standard control strategies to allow high-speed flight. However, Boeing has a patent-pending strategy that offers a solution whereby using APAS to command force (rather than position) profiles to command airplane control strategies while the helicopter swashplate positions continue to drive the cockpit inceptor positions throughout the high speed flight regimes [Ref. 32]. Table 6 summarizes the above discussion.

Table 6. Scoring of Candidate VMS Architectures with respect to High Speed Operation

	Candidate					
Criteria	I	II	III	IV.1	IV.2	IV.3
High Speed Operations	***	*+	**+	**	***	***

Reliability

Traditional FBW, because it inherently needs to be triplex, provides all the SOF and Mission reliability needed for all requirements. APAS, being duplex, is less reliable than traditional FBW from a mission reliability perspective. However because it has mechanical backup its SOF reliability is fine. APAS with coupler is similar to APAS alone since the coupler only allows more FBW features without improving reliability. Dissimilar Redundancy Extensible FBW, while triplex, operates only with either one lane active and two backups, or two lanes active and one backup, depending on the stage and design choice. Broadly speaking, the first stage provides worse mission reliability than APAS whereas the second stage provides higher mission reliability than APAS but less than a full triplex system like Traditional FBW. The third stage, while further improving mission reliability, will still be a little lower than Traditional FBW. The Dissimilar Redundancy Extensible FBW has adequate SOF reliability for all its stages. Table 7 summarizes the above discussion.

Table 7. Scoring of Candidate VMS Architectures with respect to Reliability

	Candidate					
Criteria	I	II	III	IV.1	IV.2	IV.3
Reliability	***	**	**	*	**+	***-

ARCHITECTURE IMPLICATIONS ON COST

In this section we consider the various forms of "cost" associated with each architecture. The costs considered are:

- Design Cost
- Verification and Validation (V&V) Cost
- Recurring Cost
- Operating and Sustainment (O&S) Cost
- Weight
- Space
- Risk
- Training
- Time/Schedule

In addition to qualitatively discussing the cost of each candidate architecture, we also assign a rough measure to each as was done in the section assessing the architectures' abilities to meet the requirements. We use the same scale of 1 to 3 because that is the appropriate resolution based on there being only preliminary analysis to date. Per the previous section, the scoring should be used only in the context of trends (i.e. is one architecture clearly dominating the other architectures, how do they seem to stack up against one another) and indications of where the authors' concerns lie. Future work should refine these scores with more rigorous analysis.

Design Cost

Traditional FBW's design cost is first discussed since that is the most well known. Historically, FBW design cost has proven to be very expensive. That said, companies that have done it before can really lower the design cost so this number can really vary depending on company experience and the ability to leverage that experience. For this analysis, we will still assume it is very expensive. APAS is a lot simpler than Traditional FBW to design because it only needs to be duplex and there is the option to revert back to a mechanical system. APAS with coupler design cost is somewhere in between APAS and traditional FBW, but closer to APAS. Dissimilar FBW design cost varies depending on what stage and the specific approaches taken. Its initial stage is expected to be closer, if not slightly less than, APAS costs because of its simplicity. It is expected to reach the second stage less expensively than traditional FBW. However, ultimately, because of the incremental approach not being quite as efficient as doing it all at once, the cost to the final stage is expected to be more expensive than traditional FBW. Table 8 summarizes this analysis.

Table 8. Scoring of Candidate VMS Architectures with respect to Design Cost

	Candidate					
Criteria	I	II	III	IV.1	IV.2	IV.3
Design Cost	$$$	$$	$$+	$$-	$$$-	$$$+

Verification and Validation Cost

The Verification and Validation (V&V) Cost differences between architectures are dominated by Formal Qualification Testing (FQT). FQT is an often overlooked aspect to design since it is required and treated as invariant to the design. In reality, FQT heavily depends on the design and it can be a major cost and schedule sink that has killed major aircraft programs [Ref 33]. It must be demonstrated to the certification authority and the customer that the control system meets the agreed upon requirements for contractual fulfillment. Oftentimes, designers and architects with little FQT experience tend to disregard this last mile obstacle and make choices that optimize performance and capability requirements but do not consider those impacts on certification. We examine the various candidate architectures to see whether any have a propensity to be an extremely complicated or particularly efficient FQT.

A traditional FBW system (I) with its triplex flight computer architecture and no mechanical control backup mechanism requires the most FQT because the number of cases and logic paths that must be tested goes up exponentially with overall system complexity and traditional FBW is the most complex. One of the advantages of the Dissimilar Redundancy architecture (IV) is that it reduces this complexity by breaking the system in its initial stage to a single strand FBW system, then later duplex with reversion to simpler simple FBW, before ultimately becoming triplex. By slowly building up to a full FBW system the amount of FQT required is judiciously metered out at each stage, allowing the cost of such qualification to be spread out over multiple funding cycles. Architectures (II) and (III) have V&V benefits because they preserve the original baseline mechanical flight control paths of the legacy AH-64 and also do not require a full triplex system. As such, they are both projected to have significantly less V&V cost than traditional FBW, though III will be somewhat more costly than II because it is a bit more complex. Table 9 summarizes.

Table 9. Scoring of Candidate VMS Architectures with respect to V&V Cost

	Candidate					
Criteria	I	II	III	IV.1	IV.2	IV.3
V&V Cost	$$$	$$-	$$	$	$$	$$$-

Recurring Cost

Recurring cost is the per vehicle cost due to all the additional hardware required (as well as cost to remove current hardware). This cost is by far the highest for Traditional FBW (I). It requires brand new triplex computers, some form of triplex inertial (hopefully leveraging the existing duplex EGIs), new triplex actuator remote interface units, FBW actuators, active inceptors, additional changes to the hydraulics and electrical power etc. In contrast, APAS (II) is very inexpensive and APAS+Coupler (III) is marginally more expensive. Both require active inceptors, the associated electronics, and a duplexing of the flight control computer, albeit the second flight computer can be simpler than the current AH-64 FMC since it does not need to support BUCS or SCAS operation. The Dissimilar Redundancy Stage 1 has the least cost since it involves the least new hardware and no active inceptors. The cost increases for the next two stages but falls short of that for Traditional FBW. Table 10 summarizes.

Table 10. Scoring of Candidate VMS Architectures with respect to Recurring Cost

	Candidate					
Criteria	I	II	III	IV.1	IV.2	IV.3
Recurring Cost	$$$	$+	$$-	$	$$-	$$

Operating and Sustainment Cost

Operating and Sustainment (O&S) Cost is the cost to maintain and operate the vehicle. This includes troubleshooting failures, the cost of the hardware replaced etc. When it comes to detecting and diagnosing failures, all modern digital control systems are designed to do well on this front, thus there is little distinction in that respect between the architectures. However, a simplex system lends itself to having slightly more trouble diagnosing faults than redundant systems. On the other hand, the more redundancy there is, the more line replaceable units (LRUs) there are to fail, and the overall mean time between repair decreases linearly with number of LRUs. The other aspect to consider is the O&S costs of the components that are replaced by the architecture. While the traditional FBW replaces much of the mechanical controls, those controls are not believed to play a big factor in the O&S costs, though this assertion needs to be scrutinized. The final aspect of O&S cost is rigging. Modern digital systems all tend to make that task simpler, but more research is needed to know which of these 4 architectures is best in that regard. Thus, all other things assumed equal, the O&S cost for each of the candidate architectures tends to be dominated by their total number of electronic boxes. Table 11 reflects this, giving some credit to Traditional FBW which is expected to be better than the other triplex approach, Dissimilar Redundancy FBW.

Table 11. Scoring of Candidate VMS Architectures with respect to O&S Cost

	Candidate					
Criteria	I	II	III	IV.1	IV.2	IV.3
O&S Cost	$$-	$+	$+	$	$$	$$

Weight

Weight is another important factor that can indirectly be ascribed a cost by a dollars-per-pound-of-weight-removed sensitivity. For the AH-64, that dollar to weight sensitivity is quite high compared to a commercial jet liner or even a larger helicopter like the H-47. Adding active inceptors is common to all architectures except IV.1, thus that aspect is a wash for all but IV.1. Traditional FBW with its need to be triplex redundant in flight computers, actuator remote interface units, inertial navigation units, etc. would seem to incur the heaviest weight penalty. However, various informal weight estimates over the years have shown results ranging from a slight weight gain to a weight reduction. The weight reductions are due to removal of the lower mechanical controls and lighter solutions for the upper mechanical controls (no mixer is required). We will assume for the purposes of this study that Traditional FBW is roughly weight neutral.

In contrast, the other architectures only add weight, though the additional weight is not as big as one might think. Because the SAUs and FSAs are close to the weight of the magnetic brake and feel spring they replace, APAS's weight only increases from the addition of a second flight computer which can be made smaller and lighter than the existing AH-64's FMCO. APAS+Coupler involves adding a light-weight in-line decoupler. Dissimilar Redundancy FBW involves adding simpler, and hence smaller and lighter, analog computers and two small light-weight actuators for the initial stages, at the same time removing a mechanical control linkage. Later stages replace the analog computers with extensible digital flight control computers but, while bigger than their analog counterparts, will not be that big because most of the I/O comes in through data buses from largely already-existing I/O. The last stage adds very small MEMS IMUs which are bussed directly to the digital flight computers that were designed to account for this interface in the previous stage. Table 12 reflects these space considerations using a very high dollar to weight sensitivity.

Table 12. Scoring of Candidate VMS Architectures with respect to Weight

	Candidate					
Criteria	I	II	III	IV.1	IV.2	IV.3
Weight	$-	$+	$+	$-	$$-	$$-

Space

Space on the Apache is arguably even more critical than weight. With the assortment of various mission equipment, the aircraft's environmentally conditioned areas have very little available space. Adding even a single LRU is considered significant. The space argument follows the same reasoning as the weight arguments made above and Table 13 reflects these space considerations using a very high dollar to space sensitivity (probably one of the highest in aviation).

That is, traditional FBW, while it may be the same or lower weight, requires adding electronics in places where there is no room for electronics and so, while the overall weight may perhaps be neutral or less, it is in all the wrong places. In contrast, the other architectures takes up much less space comparatively.

Table 13. Scoring of Candidate VMS Architectures with respect to Space

	Candidate					
Criteria	I	II	III	IV.1	IV.2	IV.3
Space	$$	$	$	$+	$$-	$$-

Risk

Risk is another important factor that affects the cost of mechanizing an architecture. Traditional FBW is well understood so its risk would seem to be low. However, the struggles it has to make inroads in the rotorcraft industry and the time and cost overruns to get the aircraft into production compared to other non-FBW rotorcraft indicate it to be still somewhat risky. APAS is straightforward, does not have all the complications (and hence risk) of a true FBW system, and Boeing is particularly familiar with it because of the H-47 APAS development. The one AH-64 unique element, BUCS, certainly needs to be properly addressed and may have some challenges to work through to ensure no inadvertent integration between APAS and BUCS, but those challenges are expected to be straightforward. APAS+Coupler has some additional risk due to not having done that before, but the coupler is seemingly straightforward and designs have been explored. The one area of risk is the ability to handle transients during a re-coupling from FBW to mechanical control, but this also is seen to be very well defined and addressable. The Dissimiliar Redundancy approach is a bit more risky because of it being more novel and more complicated than option I. However, its use of very reliable and simple analog backup computers, would seem to be less risk than a triplex system with all its commensurate software. This follows the logic used for the NH-90 program. Table 14 summarizes.

Table 14. Scoring of Candidate VMS Architectures with respect to Risk

	Candidate					
Criteria	I	II	III	IV.1	IV.2	IV.3
Risk	$$	$+	$+	$$-	$$	$$

Training

Training cost is dominated by the number of different modes of operation and the quantity of varying types of failure modes. All the systems involved would have a plethora of modes ranging from highly augmented and stabilized modes (e.g., ACVH, TRC) to less augmented (but more maneuverable modes (e.g. RC, RCAH). However, in addition to that, failures will necessitate having to change modes of operation. Traditional FBW will always have the same modes of operation, being designed to be physically redundant. The Dissimilar Redundancy architecture will be similar but, in its earlier stage where it is using Analog Backup Computers, will have two fundamental modes of operation (regular FBW, and a direct electrical link FBW). This will complicate training some. APAS is even more complicated because it will have the advanced control laws notionally, but needs to handle APAS failures where the pilot is no longer flying the aircraft the same way since the force feedback will be completely different and more augmented modes, which avail themselves with APAS, will no longer be available. Further, APAS will also need to support BUCS operation. Thus, APAS is significantly more complicated from a training perspective. APAS+Coupler potentially complicates training a little more than APAS by the introduction of the coupler which increases the number of failure modes, and hence modes of operation, that might now exist. Table 15 summarizes.

Table 15. Scoring of Candidate VMS Architectures with respect to Training

	Candidate					
Criteria	I	II	III	IV.1	IV.2	IV.3
Training	$	$$+	$$+	$$-	$+	$+

Time/Schedule

The time it takes each of these technologies to be incorporated into a production solution is another important factor in the evaluation of their relative suitability. Traditional FBW has proven to be a very lengthy development and certification process. In contrast, because APAS is much simpler and it has a proven mechanical flight control system it can rely on, it can be done much quicker. APAS+Coupler straddles the two, being clearly more involved than APAS but much less involved than Traditional FBW. The Dissimilar Redundancy architecture should be much less involved than Traditional FBW for its initial stage(s) but, to get to the final stage in an incremental fashion (which we have assumed), will take more time than Traditional FBW. Table 16 summarizes.

Table 16. Scoring of Candidate VMS Architectures with respect to Time/Schedule

	Candidate					
Criteria	I	II	III	IV.1	IV.2	IV.3
Time/Schedule	$$$	$	$$	$+	$$+	$$$+

SCORING SUMMARY

Table 17 summarizes how each of the various architectures stack up in value with respect to the various requirement areas.

Table 18 summarizes how each of the various architectures stack up with respect to the various cost criteria. The two tables show values and costs for the various architectures and then roll them up to an overall value and cost score. The scores are relative subjective ratings (ranging from 1 through 3) for each criteria rather than actual metrics (e.g. $). These subjective scores are based on the engineering experiences of the authors and are not based on any formal study, let alone a rigorous one. Further, the same weights are used for each value/cost assessed. In practice, a formal study would use actual metrics and would consider the weightings between categories very carefully. However, the point of this paper is simply to show a holistic approach to evaluating various architectures from which conclusions could be drawn, rather than drawing any firm conclusions from the data presented here. That said, the results are interesting and we suspect they have an element of reality to them.

These highly caveated results show that the Architectural Candidates have values ranging from 17.3 for the Dissimilar Redundancy 1 to 30 for Traditional FBW, which is quite a wide range. The results are consistent with the authors' expectations. For example, it is no surprise that the Dissimilar Redundancy 3 nears the value of Traditional FBW but is still below it. One interesting thing to note is the value of APAS compared to Traditional FBW. Previous works have shown or made arguments that a well-designed partial authority system with tactile cueing can provide 80% of the value of a traditional FBW. Our result shows that APAS has 70% of the value of FBW but our analysis included many more metrics, the additional metrics being ones with which APAS is expected to struggle. Specifically, the previous studies have limited focus on handling qualities, workload, and performance but not on aspects like autonomy, high-speed operation and DVE operations. Thus, the APAS vs FBW results presented here seem reasonably consistent with those historical results.

The Costs table show costs ranging from 12.7 to 20. The results also seem reasonably credible to the authors. Dissimilar Redundancy 1 is the least costly and it tends to be the simplest and least involved. In contrast, Dissimilar Redundancy 3 is the most costly exceeding the cost of Traditional FBW. While Dissimilar Redundancy 3 is more costly than Traditional FBW, the costs for this were based on the assumption that Dissimilar Redundancy 3 was arrived at through the first two Dissimilar Redundancy stages. If one were to go straight to Dissimilar Redundancy 3, which we refer to as "3a", the overall cost would be reduced by 2.3 to 17.7.

Table 17. Scoring Summary of Candidate VMS Architectures Ability to Support Requirements

		Candidate					
Req Area	**Wt**	I	II	III	IV.1	IV.2	IV.3
HQ/WL	1	***	**+	***-	**	***	***
DVE Op	1	***	**	**+	**-	***-	***-
Autonomy	1	***	**	**+	*+	**+	***-
Performance	1	***	**+	***-	**	***	***
Task Effcctiveness	1	***	**+	***-	**	***	***
O&S Costs	1	***	**+	***-	**	***	***
Safety	1	***	**+	***-	**	***	***
Coordinated Operations	1	***	**	**+	*+	**+	***-
High Speed	1	***	*+	**+	**	***	***
Reliability	1	***	**	**	*	**+	***-
Sum		30	21	24.7	17.3	27.7	28.7

Table 18. Scoring Summary of Various Cost Criteria for Candidate VMS Architectures

		Candidate					
Criteria	**Wt**	I	II	III	IV.1	IV.2	IV.3
Design Cost	1	$$$	$$	$$+	$$-	$$$-	$$$+
V&V Cost	1	$$$	$$-	$$	$	$$	$$$-
Recurring Cost	1	$$$	$+	$$-	$	$$-	$$
O&S Cost	1	$$-	$+	$+	$	$$	$$
Weight	1	$-	$+	$+	$-	$$-	$$-
Space	1	$$	$	$	$+	$$-	$$-
Risk	1	$$	$+	$+	$$-	$$	$$
Training	1	$	$$+	$$+	$$-	$+	$+
Time/ Schedule	1	$$$	$	$$	$+	$$+	$$$+
Sum		19.3	12.7	15.3	11.3	17.3	20

To get a better perspective of the Value/Cost tradeoff, we plot Value versus Cost for each of the architectures in Figure 7. We include the Dissimilar Redundancy 3a case (DisRed3a) in that plot as well. The plot is very illuminating for several reasons. The plot shows a classical Pareto front (red line) with all of the architectures essentially on that front. This suggests that all the architectures are valid contenders depending on funding available, though some will better meet certain requirements.

Using the highest value/cost sensitivity as the only measure, it would appear that APAS would have the most return on investment (ROI). This is illustrated by constructing the Best Value Line (green) which is the line from the origin with highest slope that first intersects any of the architectures. The architecture it intersects, in this case APAS, represents the best value/cost.

The best value line and the Pareto front also graphically illustrate that the initial and second Dissimilar Redundancy approaches (DisRed1 and DisRed2) and APAS+Coupler (APAS+C), are closer to the best value line than Traditional FBW (Trad FBW). This suggest that they should be seriously considered and is very interesting to know given that most in industry are only familiar with Traditional FBW.

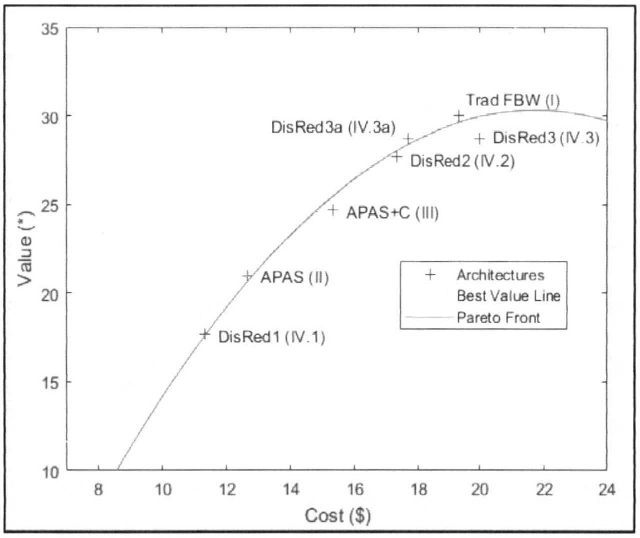

Figure 7: Value vs Cost for the Candidate Architectures

The plot also provides other insights. Specifically, the plot makes evident the benefits accrued, and the cost associated in taking incremental approaches. One incremental option is to start with Dissimilar Redundancy 1 before migrating to Dissimilar Redundancy 2. If the benefits of Dissimilar Redundancy 3 are needed, then this can be exercised. While in the end, Dissimilar Redundancy 3 will cost more than Traditional FBW, as it can be arrived at in steps as funding profiles allow. This approach is unlike Traditional FBW which must be done in one large step. This first approach has the lowest barrier to entry (if these informal results are to be believed). However, if one has sufficient budget, one can get more ROI by going to APAS first. This then has three steps, leading next to APAS+Coupler, then, we believe, to Dissimilar Redundancy 2, and finally to Dissimilar Redundancy 3. We note that the steps between each of these is much smaller than that for the first path proposed. Smaller step sizes result not only in a lower delta cost per step, but experience has shown that it minimizes risk. This small stepping concept is similar to that seen in agile software development and has been highly effective in the AH-64's avionics development.

It must be reiterated that these scores were all subjective and really more qualitative than quantitative. Other requirements and criteria might also need to considered, and no weighting was done between metrics. However, our results indicate that this methodology is informative and we think this methodology would be very useful for doing more rigorous quantitative studies on flight control system architectures. It is expected that such a methodology might also apply to comparing approaches in other technology areas as well.

One of the recognized possible difficulties with this analysis is that it boils down all cost and value scores into one lumped cost score and one lumped value score by using weightings. Thus, the approach presented does not prioritize requirements or cost criteria. Future research should investigate how to incorporate prioritization into this approach.

CONCLUSIONS

With many emerging requirements for the AH-64 attack helicopter, several VMS architectures were proposed and analyzed. The candidate VMS architectures were first defined to the extent needed to characterize their suitability against the anticipated future requirements. The architectures were then informally analyzed against those requirements and qualitative-based scores were assigned based on the authors' experiences. The architectures were also informally assessed with respect to a variety of cost metrics. A methodology was then presented which aggregated both the individual values and costs so that an overall value could be plotted against an overall cost for each architecture.

The plotted results showed that mostly all the various candidates fell along the same Pareto front. Realizing that the scores were informal and subjective, and that the same weighting was used for all metrics, it was interesting to note that the APAS architecture had the highest value/cost ratio and traditional FBW had the lowest. The plot also showed the value of incremental approaches, noting that the value/cost ratio for incremental additions, while perhaps lower than APAS, were not substantially lower. This suggests that incremental approaches should be given serious consideration.

Author contact:

Bryan Chu bryan.c.chu@boeing.com
Gary Klein gary.d.klein@boeing.com
Russell Enns: russell.enns@boeing.com
Tom Gollem: tom.gollem@boeing.com

ACKNOWLEDGMENTS

The authors would like to acknowledge the Boeing Company's significant investment in flight controls research over the years and the many ideas generated by the Mesa Fire & Flight Controls group in Mesa, AZ and the Flying Qualities Group in Philadelphia, PA. Specifically, we acknowledge the ingenious ideas from Dave Mossman and Jerry Davies of how to use existing components to provide a more affordable FBW solution for the AH-64. The authors also thank the Philadelphia Flying Qualities group and BAE Rochester UK active inceptor team for their APAS and tactile cueing technology expertise.

REFERENCES

1. Stiles, L., Mayo, J., Fresisner, A., Landis, K., & Kothmann, B., "Impossible to Resist: The Development of Rotorcraft Fly-By-Wire Technology," American Helicopter Society 60th Annual Forum Proceedings, Baltimore, MD, June 2004.
2. Tischler, M., Fletcher, J., Morris, P., & Tucker, G., "Application of Flight Control System Methods to An Advanced Combat Rotorcraft," NASA Technical Memorandum, August 1989.
3. Dones, F., Dryfoos, J., McCorvey, D., & Hindson, W., "An Advanced Fly-By-Wire Flight Control System Designed for Airborne Research – Concept to Reality," American Helicopter Society 56th Annual Forum Proceedings, Virginia Beach, VA, May 2000.
4. Robertson, G., "Flight Demonstration of an Associate System: A Rotorcraft Pilot's Association Example," American Helicopter Society 56th Annual Forum Proceedings, Virginia Beach, VA, May 2000.
5. Vidal, P., "NH-90 Helicopter Fly-By-Wire Flight Control System," American Helicopter Society 53rd Annual Forum Proceedings. Virginia Beach, VA, May 1997.
6. Sobel, J., & Reynolds, R., "525 Hydraulic System Optimization for Fly-By-Wire Control," American Helicopter Society 72nd Annual Forum Proceedings, May 2016.
7. Prouty, R., "Fly-By-Wire", *Vertiflite, Volume 55, No. 1, pp.*18, Spring 2009.
8. Taylor, A., Irwin, J., "Active Parallel Actuation Subsystem, An Alternative Approach to Tactile Cueing Without Fly By Wire", American Helicopter Society 70th Annual Forum Proceedings, Montreal, Quebec, Canada, May 2014.
9. Irwin, J., Brown, B., Rich, A., Schwerke, M., & Kocher, E., "AVMS H-47 Chinook with APAS Tactile Cueing Demonstration Results," American Helicopter Society 72nd Annual Forum Proceedings, West Palm Beach, FL, May 2016.
10. Madden, C., "Army Aviation Strike Force: A New Strategic Asset for the 21st Century," US Army War College, 1998.
11. The Boeing Company, "50th Boeing AH-64D Apache Longbow Arrives at Fort Rucker," *Defense Aerospace,* https://www.defense-aerospace.com/article-view/release/5128/50th-apache-longbow-delivered-to-ft.-rucker-%28apr.24%29.html, April 2001.
12. Enns, R., Mossman, D., & Davies, J. "Affordable Fly-By-Wire System Design and Handling Quality Considerations," Rotorcraft Handling Qualities Conference, Liverpool, UK, November 2008.
13. Enns, R., Segner, D., & Dryfoos, J. "Adaptive Vehicle Management System Program Phase II Overview," American Helicopter Society 72nd Annual Forum Proceedings, West Palm Beach, FL, May 2016.
14. Anonymous, *Handling Qualities Requirements for Military Rotorcraft, ADS-33E-PRF,* Redstone Arsenal, Alabama: US Army Aviation and Missile Command. 2015.
15. Fenley, S., Graham, D., & Gideonse, R., "Degraded Visual Environment Operational Levels," American Helicopter Society *73rd* Annual Forum Proceedings, Fort Worth, TX, May 2017.
16. Klein, G, Krainski, J, Befort, K, Enns, R., "AH-64 Apache DVE Mission Workload and Handling Qualities Simulation Study," AHS Rotorcraft Handling Qualities Technical Meeting, Huntsville, AL, February 22-23, 2017.
17. Takahashi, M., et al, "Autonomous Guidance and Flight Control on a Partial-Authority Black Hawk Helicopter", AIAA Aviation Forum, Virtual Event, June 15-19, 2020. https://doi.org/10.2514/6.2020-3286
18. Kashawlic, B., Enns, R., Irwin, J., & Moore, R., "Carefree Maneuvering via Tactile Cueing Flight Demonstrations," American Helicopter Society 72nd Annual Forum Proceedings, West Palm Beach, Florida, May 2016.
19. Kashawlic, B., Enns, R., Brown, B., & Moore, R., "Maneuver and Mission Aiding via Tactile Cueing Flight Demonstrations," American Helicopter Society 72nd Annual Forum Proceedings, West Palm Beach, Florida, May 2016.
20. Whalley, M. et al, "Flight Test Results for a New Mission-Adaptive Autonomy System on the RASCAL JUH-60A Black Hawk", American Helicopter Society 72nd Annual Forum Proceedings, West Palm Beach, Florida, May 17-19, 2016.
21. Paduano, J., et al, "TALOS: An Unmanned Cargo Delivery System for Rotorcraft Landing to Unprepared Sites," American Helicopter Society 71st Annual Forum Proceedings, Virginia Beach, VA, May 2015.
22. Cerchie, D., Caldwell D., & Chu, B., "ULB Guidance, Navigation, and Control Laws and Evaluation of their Performance during Shipboard Operations," AHS International Specialists Meeting on Unmanned Rotorcraft, Mesa, AZ, January 2012.
23. Markson, R. J., "Meet the U.S. Army's New Drone Swarms," https://mindmatters.ai/2020/09/meet-the-u-s-armys-new-drone-swarms, September 2020.
24. Prouty, R and Yackle, A., "The Lockheed AH-56 Cheyenne – Lessons Learned", AIAA Aircraft Design Systems Meeting, Hilton Head, SC, Aug 24-26 1992.
25. Klein, G, Kashawlic, B., Enns, R., "Design and Flight Test of an Adaptive Vehicle Management System on an AH-64 Helicopter Demonstrator," American Helicopter Society 72nd Annual Forum Proceedings, West Palm Beach, Florida, May 2016.
26. Hoh, R., Mitchell, D., Aponso, B., Key, D., & Blanken, C., *Background Information and User's Guide (BIUG) for Handling Qualities Requirements for Military Rotorcraft,* Moffett Field, CA: AMRDEC, December 2015.

27. Irwin, J., Chu, B., Enns, R., Klein, G., & Lenander, J., "Variable Rate Damping Based Control System for Backdriven Control Input Devices," US Patent 10,908,560, issued February 2, 2021.
28. Chu, B., Irwin, J., Enns, R., Klein, G., and Lenander, J., "Variable Operator Interface for back-driving control input devices," US Patent 11,059,568, issued July 13, 2021.
29. Chung, T., "OFFensive Swarm-Enabled Tactics (OFFSET)" https://www.darpa.mil/work-with-us/offensive-swarm-enabled-tactics.
30. Miller, D., "Swarm Warning: The Future of Unmanned Aerial Systems," https://www.army.mil/article/239210/swarm_warning_the_future_of_unmanned_aerial_systems, September 2020.
31. Trevithick, J.. "The Army Has Unveiled Its Plan For Swarms of Electronic Warfare Enabled Air-Launched Drones," https://www.thedrive.com/the-war-zone/35726/the-army-has-unveiled-its-plan-for-swarms-of-electronic-warfare-enabled-air-launched-drones, August 2020.
32. Ahmad, J., Chu, B., Enns, R., & Klein, G., Parallel Actuation Control System Providing Dual Mode Operator Control Inputs for a Compound Aircraft, US Patent Application US20210061444, published March 4, 2021.
33. Ostrower, J., "*The Mitsubishi Space Jet – as we know it today – is dead,*" https://theaircurrent.com/industry-strategy/the-mitsubishi-spacejet-as-we-know-it-today-is-dead/, May 2020.

Evolving System Integration Labs in the Era of Digital Transformation: A Brief Primer on Digital Transformation and its Implications for Test Systems and SILs

William N. (Bill) Eccles
Principal Engineer, Director of Corporate Development
Bloomy
South Windsor, CT, USA

ABSTRACT

The nature of systems and product engineering is changing rapidly with the introduction and application of Digital Transformation and a toolset consisting of Model-Based Systems Engineering, the Digital Thread, and Digital Twin. Each of these tools is briefly introduced and examined, and several characteristics for test systems which are necessary for full digital transformation implementation are derived and enumerated. The progress that the author's test equipment manufacturing and design firm is making towards providing COTS products which interact with models, the digital thread, and the digital twin is reviewed. Finally, a brief case study is reviewed which shows a 30% increase in test efficiency after implementing one of the enumerated characteristics, which was made possible through the use of COTS products.

INTRODUCTION

Bloomy has been applying automated test technologies into all areas of the product lifecycle, from R&D and validation to production and MRO, for the last 30 years across commercial and military applications. Our position as external resources for numerous organizations provides a unique perspective on the state of test systems and allows us to develop systems and methods which address multiple industries' needs. After consistent observation of the mil/aero industrial complex's struggles with adopting COTS technologies or with maintaining custom software and hardware in-the-loop systems, we began an initiative starting in 2015 to infuse the HIL market with COTS solutions which address these struggles head on. These COTS solutions have enabled enhanced test efficiency and reduced system delivery times, and were key to reducing reliance on overburdened internal technical resources.

During this same period, "Digital Transformation" rose to buzzword-worthy status to the point where it was largely an over-applied term and lost its meaning. [Ref. 1] However, the systems engineering profession took Digital Transformation very seriously and has made great strides in rearchitecting systems engineering to accommodate increasingly-complex systems. The International Council on Systems Engineering (INCOSE) introduced concepts, such as model-based systems engineering, the digital thread and twin, which embody the intent of Digital Transformation. Because our customers' products are being used by organizations which are in various stages of adopting these transformational methods, we are actively adapting our systems to fully participate in a Digitally Transformed systems engineering environment.

It is reasonable to assume that accommodating Digital Transformation in test systems means merely building them in a more flexible way in order to keep up with the ever-increasing pace of the product cycle and the risks of last-minute changes—especially during validation testing which usually occurs coincident with product development. Though this assumption is true, it is only partially responsible for the necessity for test systems to be fully Digital Transformation-capable, and it certainly does not address test systems outside of the validation phase. As a provider of test systems which address all phases of a product's lifecycle, Bloomy recognizes the imperative for all test systems, from concept to sustainment, to be more than just flexible and responsive, but also to be interactive with the new initiatives of Digital Transformation-driven changes coming to systems engineering.

As Bloomy has seen in our customer base, "digital transformation" is being implemented in numerous different ways and at inconsistent rates. Based on our observations, we will describe at a fairly high level what we believe Digital Transformation means to our customers by describing INCOSE's vision of digital-transformation-enabled systems engineering. We'll then show why we believe test systems of all kinds must be able to interact with the tools and methods of the

Presented at the VFS Modern Ops. with Future Integrated VMS Technical Meeting, West Chester, PA, Aug. 19-20, 2021.

vision. We'll then show how Bloomy-developed COTS technologies address these transformational requirement. Finally, we'll describe a brief, simple example of early, but significant, improvements obtained using these technologies in a systems integration lab (SIL) for helicopter flight controls.

SYSTEM COMPLEXITY

In the beginning, systems engineering was done with paper and pencil, draftsmen and women, and with real-world system testing—risky tests placing humans at risk, especially in vehicular systems. As available technologies increased, so did system complexity. Figure 1, a graph of "thousands of lines of code" for various aircraft models, gives us one representation of just how complexity has grown in one span of 30 years.

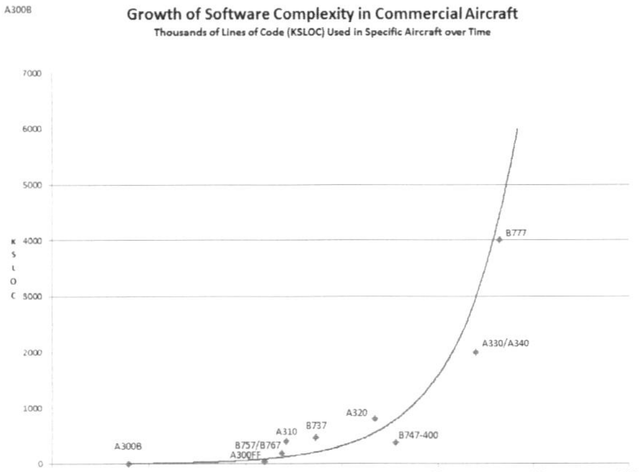

Figure 1: Thousands of Lines of Code (KSLOC) Used in Specific Aircraft over Time [Ref. 2]

One noted scholar, Christopher Alexander, has recognized that systems are too hard to predict and test.

> "Today more and more design problems are reaching insoluble levels of complexity."
>
> "At the same time that problems increase in quantity, complexity and difficulty, they also change faster than before."
>
> "Trial-and-error design is an admirable method. But it is just real-world trial and error which we are trying to replace by a symbolic method. Because trial and error is too expensive and too slow." [Ref. 3]

Given the graph of Figure 1, it would be easy to think that Alexander recognized this complexity in the mid-Nineties. However, he wrote these words in *1964,* well before software was ever found on aircraft. So it's clear that complexity was a recognized problem well before systems became as complex as they are today.

In the intervening years, of course, we've advanced to the leading edge of Industry 4.0, the age of cyber-physical systems. These cyber-physical systems consist of interconnected processes and products, and the number of possible interconnections between them has risen to the point of uncountability. A "simple" network of twenty nodes, where a node can be a physical device or a software process on a device, can be configured in 2^{190} different ways. [Ref. 4] Clearly, we can't test each of these configurations.

The number of interactions that we need to test, the configurations we need to plan for, and the nature of the systems we can build in the connected age of the "internet of things" is so large that it has eclipsed our ability to manage it. We need some method to ensure that the complexity is managed to a reasonable level, or we need something that can help us test all of the "important" interactions.

Taming Complexity with Model-Based Systems Engineering

Model-Based Systems Engineering (MBSE) can help manage the increase in system complexity. [Ref. 5]

Formally, INCOSE defines MBSE as:

> "...the formalized application of modeling to support system requirements, design, analysis, verification and validation activities beginning in the conceptual design phase and continuing throughout development and later life cycle phases." [Ref. 6]

MBSE is significantly more complex than what is presented here, [Ref. 7] but in practice, MBSE has several major characteristics.

First, it is a replacement for document-based systems engineering (DBSE), where many different distributed documents contain the characteristics, implementation, and requirements of the system. Though we've advanced from paper copies of documents to electronic copies of documents, they're still just as easy to lose and difficult to find amongst the various silos in the typical large project group. Furthermore, they are poorly revision controlled. Products such as IBM's Rational DOORS have made inroads into configuration control and traceability of requirements, tests, implementation, etc., but the data in these sorts of tools remains inconsistently expressed, making machine usage of the data an impractical task of exporting and processing before it can be used for automated purposes—a process that has to be repeated each time the data are updated.

These limitations give rise to the second and third major characteristics of MBSE: the model is expressed in a formal, machine-parsable language, System Markup Language (SysML), no matter what kind of data (whether financial, engineering, performance, etc.) is stored; and it is configuration controlled using blockchain or equivalent technologies. The last major practical characteristic of MBSE is that it is the sole, authoritative source of truth. No other repository of the data exists, thus ending countless versions of PowerPoint presentations floating around on network drives with conflicting information.

These characteristics of MBSE are enablers for other benefits. For example, the machine-parsable nature of the model allows for interaction with the model's data in previously difficult (or impossible) ways which can enhance our abilities to develop, update, maintain, and operate test systems. With proper expression of the system's interface types and quantities, the test system's design can be automatically generated if not completely, then at least in part, and can be updated quickly as requirements can, and will, change. If test requirements and limits are contained in the model, test scripts can automatically be generated and updated as well. From the perspective of a test engineer, the more data available in the model relevant to the test system and unit under test, the more opportunities there are to automate the life cycle and operations of the test systems.

The interaction between test systems and models isn't constrained to unidirectionality. A bi-directional relationship between them enables, for example, automated analysis of results, tuning of the model, predictions of reliability and trend analysis, randomized "fuzz" testing and coverage analysis, and, with the latest developments, [Ref. 8] artificial intelligence analysis of the system and data. And because of the nature of the model, all results are tied back to system requirements in the ultimate form of traceability.

Whether the test system is involved in verification, validation, production, or field support, for most efficient and effective development, maintenance and operation, the system must be able to interact with the model. Though some companies have made strides in this direction, the toolsets of MBSE are still largely unconnected to the test system lifecycle. Bloomy is working to fill this gap with our Digital Enablers for Test (DEFT™) which will include connectors between MBSE toolchains, our test equipment design processes and our customers' test scripts and methods.

Enabling Agility with Digital Twins and Digital Threads

Two other tools in the digital transformation toolkit are the digital thread and digital twin.

The connection between the model and the product's manufacturing processes was initially called the *digital thread*. The term originated at Lockheed Martin sometime prior to 2011 and "described using 3D CAD data to directly drive CNC (computer numerically controlled) milling or composite programming systems for carbon fiber placement. In both cases, the physical piece is the result of *an unbroken data link* that stretches back to the original computer model of the part; *the unbroken data path* was the digital thread." [Ref. 9] (author's emphasis)

As emphasized above, the key concept of the digital thread is the *unbroken* connection path that the data take from origin to destination. Similar to the chain of custody for evidence, the digital thread remains in tact, traceable, with no disconnects between source and destination. In practice, an updated requirement would flow through the digital thread and influence the manufacturing process appropriately.

Eventually, though, the idea of digital thread extended past the CAD data and CAM programs in both directions leading to the concept of a *digital twin*. As described on DigitalEngineering247.com, "In the [2013] USAF Global Horizons report on technology visions, digital twin is defined as 'the creation and use of a digital surrogate of a material system to allow dynamic, real-time assessment of the system's current and future capabilities.' The 'digital surrogate' is a 'physics-based technical description of the weapon system resulting from the generation, management and application of data, models and information from authoritative sources across the system's lifecycle.'" [Ref. 10] Here, for what is reportedly the first time, is a definition of a digital twin (or digital surrogate), and the definition makes clear that the digital twin involves models and all data associated with the operational and future operational characteristics of a system.

What is not made terribly clear is that the digital thread is the enabler of the digital twin, with the digital twin being a high-fidelity digital representation mirroring the lifespan of a *particular* product and *particular* serial number. The digital twin can be one or more high-fidelity computational models, combination of models, or tools which simulate and capture the life history of the related instantiation of the product. [Ref. 11]

In the same report, the USAF advocates use of a "tightly-integrated digital thread and prototyping process to enable *agile* development." The report also describes the USAF's view of the combination of the digital thread and digital twin as the "game-changer that provides the *agility* and *tailorability* needed for *rapid* development and deployment, while also reducing risk." (author's emphasis, both excerpts)

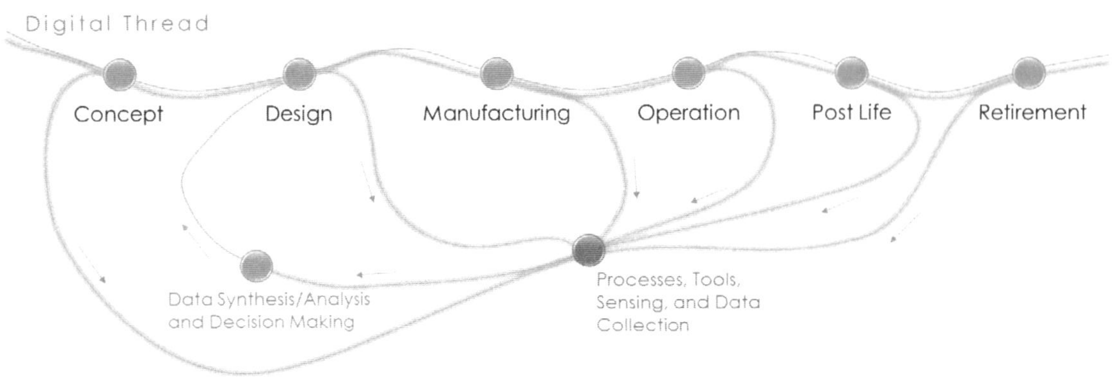

Figure 2: Illustration of engineering design with Digital Thread [Ref. 12]

Figure 2, from an academic paper which does an admirable job of expressing the mathematics of digital-thread-based design, shows an excellent representation of the way the digital thread connects both forward and backward through the product lifecycle.

Feedback from late stages can be used to influence future choices and reduce uncertainty in design parameters and process costs. This information may also reveal more efficient operational strategies, and every time these decisions are made. Each decision changes the state of the digital thread.

Taken together, the USAF's identification of digital thread/digital twin processes and the above graphic which expresses the need for the product to be able to react to both early-life changes as well as late-life discoveries, emphasizes that any test equipment used during these processes may be called upon to change as rapidly as the product may change. During early-stage iterations of the product's design, risk reduction tests using subsets of the system may be needed. But then as the product grows, so must the test system, preferably in such a way as to preserve both the investment in software and hardware made to date. This problem, which involves both hardware agility and software agility, has been solved to a large extent with Bloomy's Digital Enablers for Test (DEFT™) and the Bloomy architecture for Mil/Aero Simulation Systems, both of which are described below.

Distilled Characteristics of Digital Transformation-Ready Test Systems

In the previous sections, we've enumerated several characteristics of digital transformation which have direct implications for test systems. Distilled below are these characteristics:

1. The test system design must be able to be derived and controlled directly from the digital thread. Each design element must be traceable to an underlying requirement. Automatic, "pushbutton" test system design and manufacturing would be ideal.
2. The test system must be able to interact with the system model by retrieving tests, test limits, measurements, design parameters, methodologies, etc., from the model and digital thread. The resulting tests must be traceable to an underlying requirement.
3. Changes to the model and/or digital thread must drive changes to the test system's hardware and software. The software and hardware should be implemented in such a way that the cost of change implementation is not prohibitive.
4. Corollary: the underlying mechanisms of the hardware and software implementation should be agile and flexible in nature, easily changed quickly.
5. Data generated by test systems must be returned to the digital thread and/or digital twin and must be compatible with analysis tools which may be used to improve the system/product.
6. None of these requirements are dependent on any other, and any level of implementation of these requirements should yield an improvement to the product lifecycle.

We assume that this list is far from exhaustive, but it's at least a good start. We are also sure that this list will continue to evolve as the state of digital transformation continues to advance.

ACHIEVING DIGITAL TRANSFORMATION FOR TEST SYSTEMS: ONE COMPANY'S PROGRESS

At Bloomy, we've already tackled some of the above requirements at some level in the course of improving our own products and processes. Two products that we currently provide are used to partially address these requirements, and we're actively developing the products to further fulfil these needs. In this section, we'll describe the EFT product, which is becoming the DEFT™ product previously referenced, and the Bloomy Mil/Aero Simulation Reference System.

The EFT, originally so-named because it was designed as a software toolkit for Electronics Functional Test, provides many capabilities which Bloomy has used in our own customer deliveries for well over a decade. It grew out of our own need to meet some of the above requirements in our own development processes, and as we wrote the software, we recognized that it had intrinsic value to customers who would otherwise not purchase a test system from us and thus sell it as an independent product.

One major function which the EFT provides is hardware and measurement abstraction layers. An *abstraction layer* is a software construct which disconnects a higher-level of software from an underlying implementation, whether hardware or software. In the real world, a steering wheel is an abstraction layer because you can turn it and the vehicle's wheels turn whether the underlying steering rack is manual or power. Similarly, abstracting away the hardware in a test environment allows tests to be written irrespective of the underlying hardware. The test may be written in a hardware-agnostic fashion, making the test more portable and making it much easier to change the underlying hardware.

Similarly, a measurement abstraction layer allows the test script to contain an instruction to take a measurement without any underlying knowledge of the hardware or software required, perhaps using a set of hardware. As Figure 3 shows, the measurement abstraction layer is a higher level of abstraction than the hardware abstraction layer. It may be a complex MAL, where the underlying hardware is completely unknown, or it may be less complex and have some knowledge of the underlying hardware.

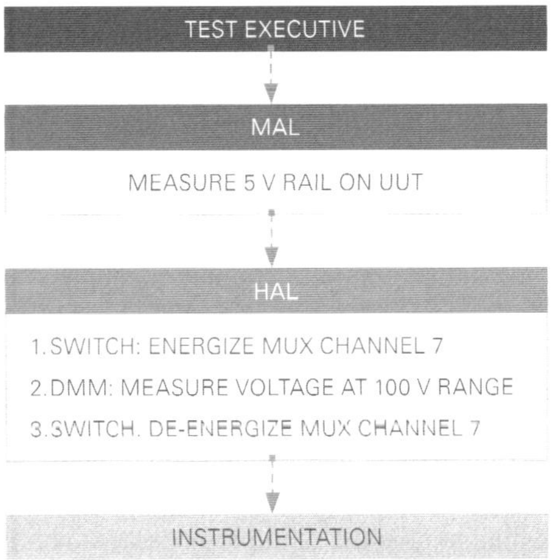

Figure 3: Abstraction Hierarchy [Ref. 13]

In this figure's example, the MAL has no idea whether a digital multimeter or an analog-to-digital converter will be used, nor does it know what switching is required to accomplish the measurement. A test script written at the HAL level would require all three steps to be explicitly written, but the same test script written at the MAL level need only know the measurement to be performed.

Both of these abstractions help attain test equipment characteristics 2-4, above. Consider that the system model in the digital thread may require "The UUT voltage must be greater than or equal to 0.5 VDC" (expressed in SysML, of course). Bloomy's MAL is able to implement the test sequences required to perform the measurement. However, a human is necessary to read the natural-language requirement and translate it into the appropriate MAL step. This is a fast process (characteristic 4), though we are working to migrate this capability to a new version of the EFT (DEFT, mentioned previously) which will be able to extract the requirement from SysML and drive the test system automatically—only one very important connection remains to be made. (Bloomy is actively seeking a lead user to work with in this capacity.) We anticipate that this step will require vendor-specific connectors for each MBSE package.

The EFT also has a plugin which addresses providing data back to a digital thread or other database (characteristic 5). It uses Structured Query Language (SQL) and a database engine connector, and for companies which use standard database systems, this capability is ready, out of the box. Again, however, where an MBSE package is being used, modification will be required to provide a vendor-specific connection to return data from the test system to the digital thread.

Similarly, we have tackled the challenges of rapid deployment of test systems by implementing a hardware design and build architecture which help address test equipment characteristics 1, 3, and 4. Twenty years ago, the state-of-the-art in test equipment system design incorporated error-prone spreadsheets, laboriously-created CAD schematics, and tedious wiring-to-print, yielding something which looked like Figure 4. Nothing was model-driven (though, to be fair, model-based design did not even exist). Changes, though possible, were certainly inconvenient and expensive. If the resources were available, several companies surveyed by the author implemented large printed circuit board assemblies which replaced the tedious wiring, but which were expensive to design and difficult to change. Again, neither design speed nor agility was a characteristic of these systems.

Figure 4: The back of a typical closed-loop test system circa 1999 [Ref. 14]

Since 2015, however, Bloomy has been working to develop the Bloomy Mil/Aero Simulation Reference System, which is modular COTS hardware and brings design and implementation up from the signal/wire level to the cable level. We have also developed our Universal Test System (UTS) platform which makes use of instrumentation assemblies which can be quickly connected together to form a complete test system from a few standard assemblies. In both platforms, COTS cables are used to connect from the mass interconnect (UTS) or Bloomy ThroughPoint™ interface panel (simulation systems), and the signal conditioning and instrumentation. This implementation allows design by *cable* instead of design by *wire*, so that systems become plug-and-play and significantly reduce construction effort and troubleshooting. This build process has allowed us to reduce system design and construction times from 6-8 months to 6-10 *weeks*.

By designing and using modular systems, we can create systems from scratch more quickly and efficiently from the preliminary I/O of a prototype product. If the product's requirements change, the system can be upgraded later as more capability is needed. This flexibility is a great feature made possible by modular, COTS hardware, but it would be useless if the software configuration of this system, which is often quite complex, weren't just as flexible and easily deployable as the hardware.

Both hardware and software flexibility and agility are required to achieve all of the above requirements. We and others have a long way to go before we've achieved total interconnection with the digital thread, but we're actively working to meet this milestone by working with our customers and implementing whatever level of thread-connectedness they are ready to achieve, as we'll see next.

DIGITAL TRANSFORMATION MEETS A SYSTEM INTEGRATION LAB

In 2016, a global air, space, and defense contractor selected the Bloomy Mil/Aero Simulation Reference System for six test systems in a helicopter system integration lab. The systems were used to verify mission systems, flight controls, engine controls, and other control systems for an airframe. The technologies used which are most relevant to digital transformation are:

- NI PXI, SLSC, VeriStand, TestStand and Requirements Gateway
- IBM Rational DOORS
- Bloomy's HAL/MAL (EFT), numerous device drivers, and other integration hardware and software.

It is worth noting that the platform is a very old airframe and, as such, "pushbutton design" was nigh impossible. Emulation of multiple layers of control logic, ranging from relay and switch logic of the 50s to intermediate levels of flight automation of the 90s, all had to be simulated to address several different airframe configurations. So test equipment characteristic 1, above, was impossible to achieve. Furthermore, the technologies associated with the Bloomy Simulation Reference System were still under development, and it took us several years to get to the point of being able to do a fairly close approximation of "pushbutton" (or scripted) design.

Because of the platform's long lifespan, many "unknowns" were discovered during the process of integrating the test systems with the units under test. Fortunately, the use of the largely-COTS Bloomy Simulation Reference System allowed for re-cabling to adjust the system to meet these previously-undiscovered requirements. Additional signal conditioning cards were added and cabled into the system, and though the system was the first of its kind ever built, the changes were made reasonably smoothly since they did not require digging through nests of wires. This "cable-level" technology was made possible because Bloomy is a test equipment company and invested millions of dollars and tens of thousands of man-hours into developing this and other test-centric products.

In rejecting the company's internal real-time test framework in favor of the COTS NI VeriStand and TestStand products, the company made two significant gains. First, no further development was required to the internal real-time test framework which has no capability to integrate with a requirements management system, much less the platform's digital thread. Given budgetary constraints usually associated with development of internal tools—namely they're usually developed with capital funds and improvements are difficult to justify since they rarely demonstrate a return on investment—and the decreasing number of personnel who are available to continue maintenance of the toolset, the minimal extra cost of the COTS software suite was well-justified.

Second, the company was able to take advantage of NI's Requirements Gateway product. The Requirements Gateway product maps requirements to test cases and coverage reports and enables test results to automatically inform the digital thread with the results of testing. Once the mapping of requirement to tests was created, test engineers were able to focus on creating test cases and making improvements to the product without worrying about requirements tracing. They were also able to input results of non-automated tests (those that required manual interaction with the system, for example) which were also mapped back to requirements. Finally, generating coverage reports was automated and prevented days of coverage analysis. The result was that test efficiency increased by *30%*.

CONCLUSION

Digital Transformation is still in its infancy, and the tools reviewed herein continue to evolve. Bloomy has witnessed a sudden uptick in the number of companies which are implementing some form of Digital Transformation. We are not, however, seeing very much consideration being given to how test systems are expected to fit into these initiatives. As a result, we expect that as more companies extend their Digital

Transformation initiatives, more emphasis will be placed on this integration. For new test systems to best be able to interact with the digital thread, digital twin, and models, COTS systems will offer the best way to achieve the test systems characteristics enumerated above, and that technologies like those offered by Bloomy will be an efficient means of achieving these goals.

Author contact: Bill Eccles Bill.Eccles@Bloomy.com

REFERENCES

1. https://techcrunch.com/2021/05/28/once-a-buzzword-digital-transformation-is-reshaping-markets/
2. https://savi.avsi.aero/about-savi/savi-motivation/exponential-system-complexity/
3. Alexander, Christopher, *Notes on the Synthesis of Form*, Harvard University Press, Cambridge Massachusetts, 1964.
4. https://www.omgwiki.org/MBSE/lib/exe/fetch.php?media=mbse:incose_mbse_iw_2020:20200126_incose_iw_mbse_transformation_v1.pdf
5. https://discover.3ds.com/sites/default/files/2020-05/simplifying-complexity-through-model-based-systems-engineering-no-magic-case-study.pdf
6. INCOSE.org
7. See INCOSE.org for more information about the other aspects of MBSE and other ways in which digital transformation is being addressed in systems engineering.
8. E.g.: https://www.ni.com/en-us/about-ni/newsroom/news-releases/ni-completes-acquisition-of-optimalplus.html
9. US Air Force, *Global Horizons Final Report: United States Air Force Global Science and Technology Vision – AF/ST TR 13-01*, United States Air Force, 2013. Excerpted by https://www.digitalengineering247.com/article/a-digital-stitch-in-time/, excerpted from
10. Ibid.
11. Singh, Victor and Willcox, Karen E., "Engineering Design with Digital Thread," *AIAA Journal* 2018 56:11, 4515-4528.
12. Ibid.
13. Gothing, Grant, *Hardware and Measurement Abstraction Layers*, National Instruments, 2016.
14. https://download.ni.com/evaluation/coretest/HAL_MAL_Web.pdf
15. Author's photo

AUTHOR INDEX